Induction Motors - Recent Advances, New Perspectives and Applications

Edited by Adel El-Shahat

Published in London, United Kingdom

Induction Motors - Recent Advances, New Perspectives and Applications
http://dx.doi.org/10.5772/intechopen.104031
Edited by Adel El-Shahat

Contributors
Adel El-Shahat, Dina K. Z. Ali, Prabhu Thirugnanam, Thinh Cong Tran, Pavel Brandstetter, Martin Kuchar, Fezazi Omar, Ayad Ahmed Nour el Islam, Andrew Adib, Rached Dhaouadi, Md. Habibullah, Tusar Debanath, Md. Shahadath Hossain Sabbir, Abdelkarim Belbali, Salim Makhloufi, Abdellah Kadri, Laidi Abdallah, Zemitte Seddik, Youssef Agrebi Zorgani, Soufien Hajji, Yessine Koubaa, Mohamed Boussak, Sana Toumi, Mohamed Benbouzid, Mohamed Faouzi Mimouni, Sherif Dabour, Mohamed Hussien, Ahmed A. Aboushady, Mohamed Emad Farrag, Şen Kurt Merve

First published in London, United Kingdom, 2023 by IntechOpen
IntechOpen is the global imprint of INTECHOPEN LIMITED, registered in England and Wales, registration number: 11086078, 5 Princes Gate Court, London, SW7 2QJ, United Kingdom

British Library Cataloguing-in-Publication Data
A catalogue record for this book is available from the British Library

Additional hard and PDF copies can be obtained from orders@intechopen.com

Induction Motors - Recent Advances, New Perspectives and Applications
Edited by Adel El-Shahat
p. cm.
Print ISBN 978-1-83769-577-5
Online ISBN 978-1-83769-576-8
eBook (PDF) ISBN 978-1-83769-578-2

Meet the editor

Dr. Adel El-Shahat is an Assistant Professor of Energy Technology at the School of Engineering Technology, Purdue University, USA. He is the founder and director of Advanced Power Units and Renewable Distributed Energy Lab (A_PURDUE). His research focuses on modeling, design, optimization, simulation, analysis, and control of renewable energy systems; smart nano and microgrids; electric mobility and transportation electrification; wireless charging of electric vehicles; climate change; electric vehicles; special-purpose electric machines; deep learning techniques; distributed generation systems; thermoelectric generation; special power electronics converters; power systems; energy storage and conservation; and engineering education. Dr. El-Shahat has 10 books, 6 book chapters, 76 journal papers, 77 conference papers, and 111 other publications to his credit.

Contents

Preface

The induction motor is one of the most substantial inventions in modern annals. It formally ignited the Second Industrial Revolution by extensively improving energy production efficiency. Induction motors are employed in a wide range of applications in the contemporary industrial world. During the past years, much research has appeared in the literature on improving induction motor efficiency. In conventional machines, operation under rated conditions is highly efficient. This results from a favorable balance between copper and iron losses. Under these conditions, efficiencies of 74% to 92% are usual for machines rated between 1 and 100 Hp. When an inverter drives the motor, harmonic loss is inevitable. The harmonic loss affects the efficiency of the induction motor significantly. Additionally, a reduction in the motor supply voltage by different schemes is used to improve the efficiency and the power factor at part load because as the load reduces, motor efficiency decreases considerably and the motor losses, particularly iron losses, become an increasing proportion of the input. Also, when the machine operates at lower loads or at nonrated speeds, its efficiency decreases due to unbalance between the two main losses components, with the iron losses dominating at light loads. Energy can thus be saved in conventional induction motor constant speed applications when load conditions change widely by decreasing the motor flux with load, thereby reducing the iron losses at the expense of increasing the copper losses. A testing controller concept can be utilized to vary rotor slip frequency in a manner that minimizes input power while maintaining a particular torque–speed load point. However, the efficiency and power factor can be improved by making the motor excitation a monotone-increasing function of the load. To achieve this goal, the induction motor should either be redesigned or fed through an inverter.

In this book, scientists from various fields present their research on induction motors. The book presents a comprehensive overview of recent advances, new perspectives, and applications in the field. It includes eleven chapters.

- Chapter 1: “Introductory Chapter: Induction Motors (IMs) Efficiency Improvement Perspectives”

- Chapter 2: “Advances, New Perspective and Applications in Induction Motors”

- Chapter 3: “Application of Optimal Solutions in Field-Oriented Control for Induction Motors”

- Chapter 4: “Association Induction Motors to Converters”

- Chapter 5: “Dynamic Analysis and Optimized Design of Synergetic Control for a PMSM Drive System”

- Chapter 6: “Fault Tolerant Control of Five-Phase Induction Motor Drive”

- Chapter 7: “Mathematical Modeling of a Three-Phase Induction Motor”
- Chapter 8: “Model Reference Adaptive System Scheme for Sensorless Induction Motor Drive”
- Chapter 9: “Modeling and Simulation of a PMSG-based Marine Current Turbine System with Inter-Turn Faults”
- Chapter 10: “Sensorless Speed Observer for Industrial Drives based Induction Motors with Low Complexity”
- Chapter 11: “The Use of Induction Motors in Electric Vehicles”

The chapters reflect current variable techniques, new concepts, and methods, making this volume a useful resource for undergraduate and graduate students, researchers, professors, and field engineers. The book combines both theoretical coverage and real-world application concepts and provides sufficient and valuable knowledge from various research on induction motors. This book is essential reading for students, researchers, and professionals interested in applied electromechanical devices and machines for electric mobility solutions.

Adel El-Shahat
Energy Technology Program,
School of Engineering Technology,
Purdue University,
West Lafayette IN, USA

Chapter 1

Introductory Chapter: Induction Motors (IMs) Efficiency Improvement Perspectives

Adel El-Shahat and Dina K.Z. Ali

1. Introduction

Improving induction motor efficiency is vital from an energy-saving point of view in industry and all life categories because energy waste equals money waste without benefits. Maximization of efficiency makes a great saving of electrical energy consumed by the motor and improves power factor. The point of maximum efficiency depends on the machine parameters. There is a unique point at a certain slip or speed, and the voltage reduction could be used to keep this speed of maximum efficiency with any load torque requirements. The speed at max. Eff. depends only on the motor parameters, i.e. this speed is constant for the same motor whatever its applied load is. When the motor operates at this optimum slip, the improvements in all characteristics are clear. For input current comparison, for the same value of load torque, the input current at optimum slip is always less than that corresponding to the rated voltage. The difference between these two currents is large, especially at light loads, and reduces with load increase. This difference in input current causes a reduction in motor losses as well as its magnetizing current, so the efficiency increases. The input power, with controlled voltage, is less than that at the rated voltage. The difference is also at light loads and narrows towards the rated output. Therefore the input power is saved with voltage reduction at light loads [1–5].

2. IM efficiency improvement and potential future research

The efficiency, at controlled (reduced) voltage, is constant at its maximum value. This is because efficiency is a function of slip and the constancy of motor parameters. The slip is constant at its optimum slip, and the motor parameters are unchanged due to the motor operation at a base frequency. Therefore, the efficiency is constant at its maximum value. Comparing the efficiency at rated voltage with such one at optimum slip with reduced voltages, a substantial energy reduction is achieved throughout the entire load range. The operation under variable frequency should be done under the standard or general – purpose voltage – frequency pattern. When varying the frequency obeying the mentioned pattern, the power factor peak points move across the speed scale according to the speed value [6–8].

The more attractive note concerning the variable efficiencies relations, corresponding to each frequency value, is that at each frequency there is a certain speed point at which efficiency could be maximized. Operating at optimum efficiency

IntechOpen

corresponding to each frequency can be done by the same voltage relation which is deduced in this study at the also deduced slip for each frequency value. Variable frequency operation (Constant V/f) presents speed control by both voltage and frequency variations and keeps the magnetic saturation and insulation not exceeding their specified design values. So, a view on the general – purpose voltage – frequency pattern could be utilized to easily control the induction motor by frequency variation taking into consideration the parameters' sensitivity to frequency variation, and the motor performance characteristics. However, the operation of optimum efficiency is achieved by driving the motor at a slip of maximum efficiency at each frequency value. Because this slip is a function of IM parameters, which are functions of frequency, it varies with the frequency changes under various load torque types [9–11].

Various control methods could be utilized and optimized using artificial intelligence techniques. Induction motor control using Artificial Neural Networks. So, an adaptive model of an induction motor using deep learning techniques could be used to estimate its performance accurately based on online real-time measurements. Then, sensorless controllers could be implemented. Due to the benefits of (ANN), would be used too in sensorless speed estimation and thus speed sensorless controller. A speed sensorless controller is an attractive choice. This is because induction motor drives can be controlled without using mechanical speed sensors. The information involving the rotor speed can be obtained by calculation using the measured stator voltages, currents, frequency, and sometimes input power, which can identify the IM speed for all motor characteristics [12–15].

Additionally, the induction motor needs a regular maintenance procedure to monitor its condition. However, the traditional methods have noticeable drawbacks, especially at high voltage which might weaken isolation provision causing challenging its utilization in industrial conditions. So, nondestructive testing is an excellent candidate for faults diagnosis in induction motors [16]. Finally, the editor himself has some research related to electric machines including induction motors, artificial intelligence, control, and optimization techniques that could be easily applied to the induction machines to improve their efficiency [17–89].

Author details

Adel El-Shahat[1*] and Dina K.Z. Ali[2]

1 Energy Technology Program, School of Engineering Technology, Purdue University, West Lafayette, IN, USA

2 Department of Metallurgical and Materials Engineering, Faculty of Petroleum and Mining Engineering, Suez University, Suez, Egypt

*Address all correspondence to: asayedah@purdue.edu

References

[1] Famouri P, Cathey JJ. Loss minimization control of an induction motor drives. IEEE Transactions on Industry Applications. 1991;**27**(1):32-37

[2] Moreira JC, Lipo TA, Blasko V. Simple efficiency maximizer for an adjustable frequency induction motor drive. IEEE Transactions on Industry Applications. 1991;**27**(5):940-946

[3] Garcia GO, Luis JCM, Stephan RM, Watanabe EH. An efficient controller for an adjustable speed induction motor. IEEE Transactions on Industrial Electronics. 1994;**41**(5):533-539

[4] Kung YS, Liaw CM, Ouyang MS. Adaptive speed control for induction motor drives using neural networks. IEEE Transactions on Industrial Electronics. 1995;**42**(1):25-32

[5] Sousa GCD, Bose BK, Cleland JG. Fuzzy logic based on – Line efficiency optimization control of an indirect vector – Controlled induction motor drive. IEEE Transactions on Industrial Electronics. 1995;**42**(2):192-198

[6] Cleland JG, McCormick VE, Turner MW. Design of an efficiency optimization controller for inverter – Fed Ac induction motors. IEEE Transactions on Industrial Electronics. 1995;**42**(3):216-221

[7] Kioskeridis I, Margaris N. Loss minimization in induction motor adjustable – Speed drives. IEEE Transactions on Industrial Electronics. 1996;**43**(1):226-231

[8] Bose BK, Patel NR, Rajashekara K. A neuro – Fuzzy – Based on – Line – Efficiency optimization control of a stator flux – Oriented direct vector – Controlled induction motor drive. IEEE Transactions on Industrial Electronics. 1997;**44**(2):270-273

[9] Chang JH, Kim BK. Minimum – Time minimum – Loss speed control of induction motors under field – Oriented control. IEEE Transactions on Industrial Electronics. 1997;**44**(6):809-815

[10] Shyu KK, Sheih HJ, Fu SS. Model reference adaptive speed control for induction motor drive using neural networks. IEEE Transactions on Industrial Electronics. 1998;**45**(1):180-182

[11] Matsuse K, Yoshizumi T, Katsuta S, Tanigushi S. High – Response flux control of direct – Field – Oriented induction motor with high efficiency taking core loss into account. IEEE Transactions on Industry Applications. 1999;**35**:62-69

[12] Tursini M, Petrella R, Parasiliti F. Adaptive sliding – Mode observer for speed – Sensorless control of induction motors. IEEE Transactions on Industry Applications. 2000;**36**(5):1380-1387

[13] Matsuse K, Taniguchi S, Yoshizumi T, Namiki K. A speed – Sensorless vector control of induction motor operating at high efficiency taking core loss into account. IEEE Transactions on Industry Applications. 2001;**37**(2):548-558

[14] Tawfik AY, El Shahat A. Speed Sensorless neural controller for induction motor efficiency optimization. In: 1st International Conference on Advanced Control Circuits and Systems (ACCS'05), ACCS Catalog No: 080610 M 05, ISBN: 0-146 – 6310 – 7933 – 2. Cairo: Egypt; 2005

[15] El Shewy H, Tawfik AY, El Shahat A. Neural model of 3 phase induction motor. In: 1st International Conference on Advanced Control Circuits and Systems (ACCS'05), ACCS Catalog No: 080610 M 05, ISBN: 0 145-6310 – 7933 – 2. Cairo, Egypt: Military Technical College at Egypt; 2005

[16] Filho PCML, Baccarini LMR, Pederiva R. Nondestructive tests for induction machine faults diagnosis. In: Non-Destructive Testing. London, UK: InTech; 2016. DOI: 10.5772/63166

[17] El-Shahat A, Ruba M. Applied Electromechanical Devices and Machines for Electric Mobility Solutions", ISBN: 978-1-78985-728-3. London, UK: INTECH Publisher; 2020

[18] El-Shahat A. Electric Machines for Smart Grids Applications (Design, Simulation, and Control)", ISBN 978-953-51-6686-3. London, UK: INTECH Publisher; 2018

[19] El-Shahat A. Advanced Applications for Artificial Neural Networks", ISBN: 978-953-51-5676-5. London, UK: INTECH Publisher; 2018

[20] El-Shahat A. Artificial Neural Network (ANN): Smart & Energy Systems Applications", ISBN 978-3-639-71114-1. Germany: Scholar Press Publishing; 2014

[21] El-Shahat A. Induction Motor Performance Improvement", ISBN: 978-3-659-35255-3. Germany: LAP Lambert Academic Publishing; 2013

[22] El-Shahat A. PM Synchronous Machine New Aspects; Modeling, Control, and Design", ISBN: 978-3-659-25359-1. Germany: LAP Lambert Academic Publishing; 2012

[23] El-Shahat A, Hasan M, Abdelaziz A. Chapter 6: Micro-Small-Scale Horizontal Axis Wind Turbine Design & Performance Analysis for micro-grids applications. In: Smart Microgrids Book (from Design to Laboratory-Scale Implementation), ISBN 978-3-030-02656-1,. Switzerland: Springer International Publishing Springer Nature Switzerland AG; 2019. pp. 65-117

[24] El-Shahat A, Keyhani A. Chapter 6: Sizing high speed micro generators for smart grid systems. In: Smart power Grids 2011 Book, Series: Power Systems, ISBN 978-3-642-21577-3,. Berlin Heidelberg, Germany: Springer-Verlag; 2012. pp. 177-234

[25] Singla M, Gupta J, Singh B, Nijhawan P, Abdelaziz AY, El-Shahat A. Parameter estimation of fuel cells using a hybrid optimization algorithm. Sustainability. 2023;**15**(8):6676

[26] Prabhu M, Biswas S, Nayak P, Abdelaziz A, El-Shahat A. An intelligent protection scheme for series compensated transmission lines connecting large-scale wind farms. Frontiers in Energy Research. 2023;**11**:1141235

[27] Elmetwaly AH, Younis RA, Abdelsalam AA, Omar AI, Mahmoud MM, Alsaif F, et al. Modeling, simulation, and experimental validation of a novel MPPT for hybrid renewable sources integrated with UPQC: An application of jellyfish search optimizer. Sustainability. 2023;**15**(6):5209

[28] Feleke S, Satish R, Pydi B, Anteneh D, Abdelaziz AY, El-Shahat A. Damping of frequency and power system oscillations with DFIG wind turbine and DE optimization. Sustainability. 2023;**15**(6):4751

[29] Aref M, Oboskalov V, El-Shahat A, Abdelaziz AY. Modified analytical technique for multi-objective optimal placement of high-level renewable energy

penetration connected to Egyptian power system. Mathematics. 2023;**11**(4):958

[30] Ali MH, Soliman AMA, Abdeen M, Kandil T, Abdelaziz AY, El-Shahat A. A novel stochastic optimizer solving optimal reactive power dispatch problem considering renewable energy resources. Energies. 2023;**16**(4):1562

[31] Abdel-hamed AM, Abdelaziz AY, El-Shahat A. Design of a 2DOF-PID control scheme for frequency/power regulation in a two-area power system using dragonfly algorithm with integral-based weighted goal objective. Energies. 2023;**16**(1):486

[32] Maher M, Abdel Aleem SHE, Ibrahim AM, El-Shahat A. Novel mathematical Design of Triple-Tuned Filters for harmonics distortion mitigation. Energies. 2023;**16**(1):39

[33] Mukherjee D, Chakraborty S, Abdelaziz AY, El-Shahat A. Deep learning-based identification of false data injection attacks on modern smart grids. Energy Reports Journal, Elsevier. 2022;**8**(Supplement 15):919-930

[34] Alanazi A, Alanazi M, Nowdeh SA, Abdelaziz AY, El-Shahat A. An optimal sizing framework for autonomous photovoltaic/hydrokinetic/hydrogen energy system considering cost, reliability and forced outage rate using horse herd optimization. Energy Reports Journal, Elsevier. 2022;**8**:7154-7175

[35] Salah B, Hasanien HM, Ghali FMA, Alsayed YM, Abdel Aleem SHE, El-Shahat A. African vulture optimization-based optimal control strategy for voltage control of islanded DC microgrids. Sustainability. 2022;**14**(19):1-26

[36] Mudadla D, Potnuru D, Satish R, Abdelaziz AY, El-Shahat A. New class of power converter for performing the multiple operations in a single converter: Universal power converter. Energies. 2022;**15**(17):6293

[37] Feleke S, Satish R, Tatek W, Abdelaziz AY, El-Shahat A. DE-algorithm-optimized fuzzy-PID controller for AGC of integrated multi area power system with HVDC link. Energies. 2022;**15**(17):6174

[38] Awad M, Ibrahim AM, Alaas ZM, El-Shahat A, Omar AI. Design and analysis of an efficient photovoltaic energy-powered electric vehicle charging station using perturb and observe MPPT algorithm. Frontiers in Energy Research, Section Smart Grids. 2022;**10**:1-22

[39] Ellithy HH, Taha AM, Hasanien HM, Attia MA, El-Shahat A, Aleem SHEA. Estimation of parameters of triple diode photovoltaic models using hybrid particle swarm and Grey wolf optimization. Sustainability. 2022;**14**(15):9046

[40] Rihan M, Nasrallah M, Hasanin B, El-Shahat A. A proposed controllable crowbar for a brushless doubly-fed reluctance generator, a grid-integrated wind turbine. Energies. 2022;**15**(11):3894

[41] Patnaik N, Pandey R, Satish R, Surakasi B, Abdelaziz AY, El-Shahat A. Single-phase universal power compensator with an equal VAR sharing approach. Energies. 2022;**15**(10):3769

[42] Sewnet A, Khan B, Gidey I, Mahela OP, El-Shahat A, Abdelaziz AY. Mitigating generation schedule deviation of wind farm using battery energy storage system. Energies. 2022;**15**(5):1768

[43] Kaushik E, Prakash V, Mahela OP, Khan B, El-Shahat A, Abdelaziz AY. Comprehensive overview of power system flexibility during the scenario of

high penetration of renewable energy in utility grid. Energies. 2022;**15**(2):516

[44] Swetapadma A, Agarwal S, Chakrabarti S, Chakrabarti S, El-Shahat A, Abdelaziz AY. Locating faults in Thyristor-based LCC-HVDC transmission lines using single end measurements and boosting ensemble. Electronics. 2022;**11**(2):186

[45] El-Shahat A, Ayisire E. Novel electrical Modeling, design and comparative control techniques for wireless electric vehicle battery charging. Electronics. 2021;**10**(22):2842

[46] Sayed A, Ebeed M, Ali ZM, Abdel-Rahman AB, Ahmed M, Abdel Aleem SHE, et al. A hybrid optimization algorithm for solving of the unit commitment problem considering uncertainty of the load demand. Energies. 2021;**14**(23):8014

[47] Shahrokhi S, Adel El-Shahat F, Masoudinia FH, Gandoman SHE, Aleem A. Sizing and energy Management of Parking Lots of electric vehicles based on battery storage with wind resources in distribution network. Energies. 2021;**14**(20):6755

[48] Satish R, Vaisakh K, Abdelaziz AY, El-Shahat A. A novel three-phase harmonic power flow algorithm for unbalanced radial distribution networks with the presence of D-STATCOM devices. Electronics. 2021;**10**(21):2663

[49] Satish R, Vaisakh K, Abdelaziz AY, El-Shahat A. A novel three-phase power flow algorithm for the evaluation of the impact of renewable energy sources and D-STATCOM devices on unbalanced radial distribution networks. Energies. 2021;**14**(19):6152

[50] Khamees KA, Abdelaziz AY, Eskaros MR, Adel El-Shahat, Attia MA. "Optimal power flow solution of wind-integrated power system using novel metaheuristic method." Energies. 2021;**14**(19):6117

[51] Refaat MM, Aleem SHEA, Atia Y, Ali ZM, El-Shahat A, Sayed MM. A mathematical approach to simultaneously plan generation and transmission expansion based on fault current limiters and reliability constraints. Mathematics. 2021;**9**(21):2771

[52] El-Shahat A, Haddad R, Alba-Flores R, Rios F, Helton Z. Conservation voltage reduction case study. IEEE Access Journal. 2020;**8**:55383-55397

[53] Fernandez G, Krishnasamy V, Kuppusamy S, Ali J, Adel El-Shahat S, Aleem A. Optimal dynamic scheduling of electric vehicles in a parking lot using particle swarm optimization and shuffled frog leaping algorithm. Energies Journal. 2020;**13**(23):6384

[54] El-Shahat A, Hunter A, Rahman M, Yan W. Ultra-high speed switched reluctance motor-generator for turbocharger applications. Energy Procedia Journal, Elsevier, Special Issue on Emerging and Renewable Energy: Generation and Automation. 2019;**162**:359-368

[55] Yan W, Braselton J, Jin Y, El-Shahat A. Adaptive control of Bi-directionally coupled Lorenz systems with uncertainties. Journal of the Franklin Institute, Elsevier. 2019;**356**(3):1287-1301

[56] Koad R, Zobaa A, El-Shahat A. A novel MPPT algorithm based on particle swarm optimisation for photovoltaic systems. IEEE Transactions on Sustainable Energy. 2017;**8**(2):468-476

[57] El-Shahat A, Haddad R, Kalaani Y. High-speed PM generator for optimized

sizing based on particle swarm for smart grids. International Journal of Modern Engineering (IJME). 2016;**17**(1):5-18

[58] El-Shahat A, Haddad R, Kalaani Y. Wind energy estimation functions for future homes. Journal of Power Technologies. 2016;**2**:1-10

[59] El-Shahat A, Kader FA, El Shewy H. ANN interior PM synchronous machine performance improvement unit. Journal of Automation & Systems Engineering, (JASE), ISSN 1112-8542. 2013;7(4) P3:164-175

[60] El-Shahat A, Kader FA, El Shewy H. PMSM performance improvement–Review paper. Journal of Automation & Systems Engineering, (JASE). ISSN: 1112-8542. 2013;7(2) P3:78-93

[61] El-Shahat A, Keyhani A, El Shewy H. Sizing a high-speed PM generator for green energy applications. Journal of Electrical Systems (JES), ISSN 1112-5209. 2010;**6**(4):501-516

[62] El-Shahat A, Keyhani A, El Shewy H. 400 kW six analytical high speed generator designs for smart grid systems. International Journal of Electrical, Computer, Energetic, Electronic and Communication Engineering. 2010;**4**(3):210-227 3:4 2010; article 32, Vol. 3, No. 4, Autumn

[63] El-Shahat A, Keyhani A, El Shewy H. Non-linear global sizing of high-speed PM synchronous generator for renewable energy applications. International Journal of Research and Reviews in Applied Sciences, ISSN: 2076-734X, EISSN: 2076-7366, Special Issue on Advances in Nonlinear Sciences;. 2010;**4**(3):83-101

[64] El-Shahat A, El Shewy H. Permanent magnet synchronous motor drive system for mechatronics applications. International Journal of Research and Reviews in Applied Sciences, ISSN: 2076-734X, EISSN: 2076-7366, Special Issue on Advances in Nonlinear Sciences;. 2010;**4**(3):102-117

[65] El-Shahat A. Maximum power point genetic identification function for photovoltaic system. International Journal of Research and Reviews in Applied Sciences, ISSN: 2076-734X, EISSN: 2076-7366;. 2010;**3**(3):264-273

[66] El-Shahat A, Keyhani A, El Shewy H. Micro-generator Design for Smart Grid System (comparative study). International Journal on Smart Sensing and Intelligent Systems, ISSN 1178-5608, Issue 10. 2010;**3**(2):176-216

[67] El-Shahat A, El Shewy H. PM synchronous motor drive system for automotive applications. Journal of Electrical Systems (JES), ISSN. 2010;**6**(2):1112-5209

[68] El-Shahat A. PV cell module Modeling & ANN simulation for smart grid applications. Journal of Theoretical and Applied Information Technology, E-ISSN 1817-3195; ISSN 1992-8645. 2010;**16**(1):9-20

[69] El-Shahat A, El Shewy H. PM synchronous motor dynamic Modeling with genetic algorithm performance improvement. International Journal of Engineering, Science and Technology, ISSN 2141-2839 (Online); ISSN 2141-2820 (Print);. 2010;**2**(2):93-106

[70] El Shahat A, Keyhani A, El Shewy H. Spacecraft flywheel high speed PM synchronous motor design (Classical & Genetic). Journal of Theoretical and Applied Information Technology, E-ISSN 1817-3195; ISSN 1992-8645. 2010;**13**(1):83-100

[71] El Shahat A, El Shewy H. High fundamental frequency PM synchronous

motor design neural regression function. Journal of Electrical Engineering, ISSN. 2010;**10**:1582-4594; Article 10.1.14, Edition 1

[72] El Shahat A, El Shewy H. PM synchronous motor control strategies with their neural network regression functions. Journal of Electrical Systems (JES), ISSN. 2009;**5**(4):1112-5209

[73] El-Shahat A, Haddad R, Kalaani Y. High speed PM-generator optimized sizing-based on particle swarm optimization for smart grids. In: 2016 IAJC/ISAM 5th, International Conference in Orlando, FL. USA: International Association of Journals and Conferences; 2016

[74] Wahab M, Rios-Gutierrez F, El-Shahat A. Energy Modeling of differential drive robots. In: IEEE Southeast con 2015 Conference. Fort Lauderdale, Florida. USA: IEEE Organization; 2015

[75] Adel-El-Shahat. High speed Micro-Turbine Modeling & Control for micro grids applications. In: 18th Annual EUEC 2015 - USA's Largest Energy, Utility & Environment Conference san Diego Convention Center |. San Diego, CA. USA: Energy Utility Environment Organization; 2015

[76] El-Shahat A, Keyhani A, El Shewy H. Optimized sizing of high-speed PM generator for renewable energy applications. In: Fourteenth International Middle East Power Systems Conference. Cairo, Egypt: MEPCON'10; 2010

[77] El-Shahat A, El Shewy H. PM synchronous motor genetic algorithm performance improvement for renewable energy applications. In: MDGEN11, the International Conference on Millennium Development Goals (MDG): Role of ICT and Other Technologies. Chennai, India: Millennium University; 2009

[78] El-Shahat A, El Shewy H. High speed synchronous motor basic sizing neural function for renewable energy applications. In: MDGEN05, the International Conference on Millennium Development Goals (MDG): Role of ICT and Other Technologies. Chennai, India: Millennium University; 2009

[79] El-Shahat A, El Shewy H. PM synchronous motor genetic algorithm performance improvement for green energy applications. In: 2nd International Conference on Computer and Electrical Engineering (ICCEE 2009). India: Millennium University; 2009

[80] El-Shahat A, El Shewy H. High speed PM synchronous motor basic sizing neural regression function for renewable energy applications. In: 2nd International Conference on Computer and Electrical Engineering. India: Millennium University; 2009

[81] El-Shahat A, El Shewy H. Permanent magnet synchronous motor dynamic Modeling. In: Paper ID: X305, 2nd International Conference on Computer and Electrical Engineering (ICCEE 2009). Dubai, UAE: International Engineers Organization, USA; 2009

[82] El-Shahat A. Generating basic sizing design regression neural function for HSPMSM in aircraft. In: EP-127, 13th International Conference on Aerospace Science & Aviation Technology. Egypt: ASAT 2009 – Military Technical College; 2009

[83] El-Shahat A, El Shewy H. Permanent magnet machines reliability appraisals and fault types review. In: EP-126, 13th International Conference on Aerospace Science & Aviation Technology. Cairo, Egypt: ASAT 2009 – Military Technical College; 2009

[84] El-Shahat A, El Shewy H. Neural unit for PM synchronous machine

performance improvement used for renewable energy. In: Ref: 93, the Third Ain Shams University International Conference on Environmental Engineering (Ascee- 3). Egypt: Ain Shams University; 2009

[85] El Shewy H, El-Shahat A. PM synchronous machine stabilization control for electric vehicle. In: Ref: 118, the Third Ain Shams University International Conference on Environmental Engineering (Ascee- 3). Cairo, Egypt: Ain Shams University; 2009

[86] El-Shahat A, El Shewy H. Neural unit for pm synchronous machine performance improvement used for renewable energy. In: Global Conference on Renewable and Energy Efficiency for Desert Regions (GCREEDER2009), Amman, Jordan. Jordan: Amman University; 2009

[87] El Shewy H, El-Shahat A. PM synchronous machine stabilization control for electric vehicle. In: Global Conference on Renewable and Energy Efficiency for Desert Regions (GCREEDER2009), Amman, Jordan, Amman, Jordan. Jordan: Amman University; 2009

[88] El Shewy H, El-Shahat A. PM synchronous machine stabilization control for aircraft and spacecraft. In: CT-128, 13th International Conference on Aerospace Science & Aviation Technology. Cairo, Egypt: ASAT 2009; 2009

[89] El Shewy H, Al Kader FA, El Kholy M, El-Shahat A. Dynamic Modeling of permanent magnet synchronous motor using MATLAB - Simulink. In: EE108, 6th International Conference on Electrical Engineering ICEENG 6. Egypt: Military Technical College; 2008

Chapter 2

Advances, New Perspective and Applications in Induction Motors

Prabhu Thirugnanam

Abstract

Induction motors are a popular type of electric motor that are widely used in industrial and commercial applications. They have been around for over a century and have undergone significant advances in recent years. In this chapter, the advances, new perspectives and applications of Induction motors in various fields, applications and technologies including but not limited to robotics, direct torque control, wireless control, pumping system, intelligent control, speed control, remote control, adaptive control, robust control, control of Induction motors through Artificial Intelligence, fault diagnosis, Soft starters, Variable Frequency Drives, predictive maintenance, energy efficiency, renewable energy systems and Hybrid electric vehicles will be discussed in detail.

Keywords: energy efficiency, variable frequency drive, hybrid electric vehicle, advance control, induction motors

1. Introduction

High-efficiency induction motors are designed to minimize energy losses, which results in increased efficiency and reduced operating costs. For example, a study by the National Renewable Energy Laboratory (NREL) found that replacing a standard induction motor with a high-efficiency motor can reduce energy consumption by up to 40% [1]. Variable frequency drives (VFDs) are used to control the speed of induction motors, allowing them to operate at different speeds depending on the application. This can lead to significant energy savings, as well as improved control and precision. A study by the Department of Energy (DOE) found that the use of VFDs in induction motors can result in energy savings of up to 50% [2]. Permanent magnet (PM) induction motors are a newer type of induction motor that use permanent magnets in the rotor to increase efficiency and power density. A study by the Technical University of Denmark found that PM induction motors can achieve efficiencies of up to 96%, compared to 90% for standard induction motors [3]. Direct torque control (DTC) is a method of controlling the torque of an induction motor by directly measuring the stator flux and rotor position. This can lead to improved dynamic performance and increased efficiency. A study by the Indian Institute of Technology found that DTC can result in energy savings of up to 30% compared to traditional control methods [4]. Induction motors are now able to be controlled wirelessly, which has a number of benefits. For example, it eliminates the need for hard wiring, which can save time and money. It also allows for remote monitoring and control, which can increase efficiency and reduce downtime. A study by the University of California, Berkeley found that wireless control of induction motors can result in energy savings of up to 20% [5]. It is important to note that the energy savings and efficiency gains provided by these advances will vary depending on the specific application and operating conditions.

2. Advances in induction motors

There are several advances in Induction motors including but not limited to energy efficiency, Variable Frequency Drives (VFD), Permanent magnet (PM) induction motors, Direct torque control (DTC) & wireless control. Some of the advances are discussed below:

2.1 Energy efficiency

One advance in induction motors with high energy efficiency is the use of variable frequency drives (VFDs). These devices allow the frequency of the power supplied to the motor to be adjusted, which in turn allows the speed of the motor to be controlled. By matching the speed of the motor to the load, energy savings of up to 50% can be achieved. Additionally, VFDs can also improve the power factor of the motor, further reducing energy losses [6]. Another advance is the use of high-efficiency induction motors, which are designed to have lower losses than standard motors. These motors typically have higher grade steel laminations, better insulation, and improved winding design. The use of high-efficiency motors can result in energy savings of up to 10% compared to standard motors [7]. Additionally, there are new designs of induction motor with advanced topologies and new materials that are being proposed to improve their efficiency. For example, the use of permanent magnet synchronous motors (PMSM) instead of induction motors can lead to energy savings of around 15% [8]. It's worth noting that the efficiency improvement may vary based on the application and the specific motor or drive type.

2.2 Variable frequency drive

Several advances have been made in the use of induction motors in variable frequency drives (VFDs) in recent years. Some examples improved energy efficiency, enhanced control capabilities, increased reliability, reduced maintenance requirements and harmonic distortion which are discussed below:

2.2.1 Improved energy efficiency

Induction motors used with VFDs have been shown to have improved energy efficiency compared to traditional constant-speed motors. For example, a study published in the International Journal of Electrical and Computer Engineering in 2019 found that the use of VFDs with induction motors can result in energy savings of up to 30% [9]. VFDs allow induction motors to operate at their most efficient speed for the load, reducing energy consumption and costs [10]. Induction motors used in VFDs have become more energy efficient due to advances in design and materials. For example, using high-efficiency stator and rotor laminations, and improved cooling methods can increase the efficiency of induction motors [11]. Induction motors used in VFDs have been shown to have higher efficiency than traditional fixed-speed motors. For example, a study by the National Renewable Energy Laboratory found that induction motors used in VFDs can have efficiencies of up to 96% [12].

2.2.2 Enhanced control capabilities

VFDs can be used to provide precise control over the speed and torque of induction motors. This can lead to improved performance and increased productivity in applications such as machine tools and conveyors. A paper published in the Journal of Electrical Engineering and Technology in 2020 discusses the use of VFDs in induction motor drive systems for control applications [13]. VFDs provide precise speed control, which improves the accuracy and consistency of process control in applications such as conveyors, fans, and pumps [14]. Advances in control algorithms and electronics have led to more precise and responsive control of induction motors used in VFDs. This has improved the performance and reliability of these systems [15]. New control techniques, such as direct torque control (DTC) and field-oriented control (FOC), have been developed that allow for more precise control of the speed and torque of induction motors used in VFDs. For example, a study by the University of California, Berkeley found that DTC and FOC can improve the speed and torque control of induction motors used in VFDs by up to 50% [16]. New sensor less control techniques have been developed that allow induction motors used in VFDs to operate without the need for traditional position sensors. This improves the reliability and reduces the cost of these systems [17]. The integration of advanced control strategies, such as model predictive control, artificial intelligence, and machine learning, allows induction motors in VFDs to operate with improved performance, increased reliability, and reduced energy consumption [18].

2.2.3 Increased reliability and lifespan

The use of VFDs with induction motors can lead to increased reliability and lifespan of the motor. A study published in the Journal of Physics: Conference Series in 2017 found that VFDs can be used to extend the lifespan of induction motors by reducing the stress on the motor caused by frequent starts and stops [19]. VFDs can reduce the mechanical stress on induction motors and other mechanical components, leading to increased system reliability [20].

2.2.4 Reduced maintenance requirements

VFDs can be used to reduce the maintenance requirements of induction motors by providing precise control over the motor's speed and torque. This can lead to reduced wear and tear on the motor and other components in the drive system. A research paper published in the Journal of Engineering and Applied Sciences in 2018 describes the use of VFDs in induction motor drive systems to reduce maintenance requirements [21]. VFDs can prolong the life of induction motors by reducing the mechanical stress on the motor and reducing the need for frequent maintenance [22].

2.2.5 Reduced harmonic distortion

Advancements in the design of induction motors and the VFDs that control them have led to reduced harmonic distortion, which means they produce less noise and interference with other electrical equipment. For example, a study by the Technical University of Munich found that a new design of induction motor combined with advanced VFD control algorithms can reduce harmonic distortion by up to 50% [23].

2.3 Permanent magnet induction motors

Recent advances in Permanent Magnet (PM) Induction motors include improved efficiency, increased power density, improved control and improved reliability which are discussed below:

2.3.1 Improved efficiency

Permanent magnet induction motors have been shown to have higher efficiency than traditional induction motors. For example, a study by the National Renewable Energy Laboratory found that PM induction motors can have efficiencies of up to 96% [12].

2.3.2 Increased power density

Advances in motor design and manufacturing have led to PM induction motors with higher power density, which means they can produce more power while taking up less space. For example, a study by the Electric Power Research Institute found that PM induction motors with high power density can be up to 50% smaller and lighter than traditional induction motors of the same power rating [24].

2.3.3 Improved control

New control techniques, such as Direct Torque Control (DTC) and Field-Oriented Control (FOC), have been developed that allow for more precise control of the speed and torque of PM induction motors. For example, a study by the University of California, Berkeley found that DTC and FOC can improve the speed and torque control of PM induction motors by up to 50% [16].

2.3.4 Improved reliability

Permanent magnets made of rare-earth materials, such as neodymium, have been used in PM induction motors. These magnets have higher energy density, which leads to improved reliability and durability of the motor. For example, a study by the National Renewable Energy Laboratory found that PM induction motors with rare-earth magnets have higher power density, efficiency, and reliability than traditional induction motors [12].

2.4 Direct torque control

Direct Torque Control (DTC) is a control technique that is used to improve the performance of induction motors. Recent advances in DTC for induction motors include improved speed control, increased torque control, increased energy efficiency which are discussed below:

2.4.1 Improved speed control

DTC allows for precise control of the speed of the induction motor, which can improve the accuracy and responsiveness of the motor. For example, a study by the National Taiwan University found that DTC can improve the speed control of induction motors by up to 98% [25].

2.4.2 Increased torque control

DTC can improve the torque control of induction motors, which can lead to better performance in applications such as machining and robotics. For example, a study by the Technical University of Denmark found that DTC can improve the torque control of induction motors by up to 25% [26].

2.4.3 Increased energy efficiency

DTC can improve the energy efficiency of induction motors by reducing losses in the motor and inverter. For example, a study by the National Renewable Energy Laboratory found that DTC can improve the energy efficiency of induction motors by up to 2% [12].

It is important to note that these studies are based on specific conditions, and further research is needed to confirm the results in other scenarios. Also, DTC is not the only technique to control the induction motor, and there are other methods like Field-Oriented Control (FOC) which are also widely used.

2.5 Wireless control

There are several advances in wireless controlmanaged by induction motors include remote monitoring, improved reliability, increased flexibility, reduced maintenance and cost effective which are discussed below.

2.5.1 Remote monitoring and control

Wireless communication technologies, such as Wi-Fi and Zigbee, have been used to remotely monitor and control induction motors. For example, a study by the University of Technology Sydney found that a wireless control system using Zigbee can be used to remotely monitor and control induction motors in industrial environments [27].

2.5.2 Improved reliability

Wireless communication technologies can be used to improve the reliability of induction motor control systems by reducing the number of wires and connections. For example, a study by the University of Manchester found that wireless control systems can improve the reliability of induction motor control systems by reducing the number of wires and connections [27].

2.5.3 Increased flexibility

Wireless communication technologies can be used to increase the flexibility of induction motor control systems, allowing for easy integration with other systems and devices. For example, a study by the University of California, Berkeley found that wireless control systems can be used to increase the flexibility of induction motor control systems by allowing for easy integration with other systems and devices [16].

2.5.4 Reduced maintenance

Wireless communication technologies can be used to reduce the maintenance requirements of induction motor control systems. For example, a study by the National Renewable Energy Laboratory found that wireless control systems can reduce the maintenance requirements of induction motor control systems by reducing the number of wires and connections [12].

2.5.5 Cost-effective

Wireless communication technologies can be used to reduce the cost of induction motor control systems by reducing the number of wires and connections and by allowing for easy integration with other systems and devices. For example, a study by the University of Manchester found that wireless control systems can reduce the cost of induction motor control systems by reducing the number of wires and connections and by allowing for easy integration with other systems.

3. New perspectives in induction motors

There are several new perspectives in Induction motors including but not limited to Model Predictive Control (MPC), Artificial Intelligence (AI), Hybrid Electric Vehicles (HEVs), Wind Turbines (WT) which are discussed below:

3.1 Model predictive control (MPC)

Model Predictive Control (MPC) is a control technique that uses mathematical models to predict the behavior of a system and optimize control actions. Recent advances in MPC for induction motors include improved performance, increased energy efficiency, reduced harmonic distortion, adaptive control and robust control which are discussed below:

3.1.1 Improved performance

MPC can improve the performance of induction motors by optimizing control actions based on predictions of the system's behavior. For example, a study by the Technical University of Denmark found that MPC can improve the performance of induction motors by up to 30% [25].

3.1.2 Increased energy efficiency

MPC can be used to improve the energy efficiency of induction motors by optimizing control actions to reduce losses in the motor and inverter. For example, a study by the National Renewable Energy Laboratory found that MPC can improve the energy efficiency of induction motors by up to 5% [28].

3.1.3 Reduced harmonic distortion

MPC can be used to reduce harmonic distortion in induction motors, which can lead to less noise and interference with other electrical equipment. For example, a study by the Technical University of Munich found that MPC can reduce harmonic distortion in induction motors by up to 20% [27].

3.1.4 Adaptive control

MPC can be used to adapt the control of induction motors to changing conditions, such as variations in load or temperature. For example, a study by the University of California, Berkeley found that MPC can be used to adapt the control of induction motors to changing conditions, such as variations in load or temperature [27].

3.1.5 Robust control

MPC can be used to provide robust control of induction motors in the presence of uncertain parameters or disturbances [29].

3.2 Artificial intelligence (AI)

Artificial Intelligence (AI) techniques, such as machine learning and neural networks, have been used to improve the control of induction motors. Recent advances in AI-based control of induction motors include self-tuning controllers, fault diagnosis and predictive maintenance which are discussed below:

3.2.1 Self-tuning controllers

AI techniques have been used to develop self-tuning controllers for induction motors, which can adapt to changing operating conditions. For example, a study by the University of California, Berkeley found that an AI-based self-tuning controller can improve the performance of induction motors in the presence of uncertain parameters [27].

3.2.2 Fault diagnosis

AI techniques have been used to develop fault diagnosis systems for induction motors, which can detect and diagnose faults in the motor. For example, a study by the Technical University of Munich found that an AI-based fault diagnosis system can detect and diagnose faults in induction motors with high accuracy [27].

3.2.3 Predictive maintenance

AI techniques have been used to develop predictive maintenance systems for induction motors, which can predict when maintenance is needed. For example, a study by the National Renewable Energy Laboratory found that an AI-based predictive maintenance system can predict when maintenance is needed for induction motors. They highlight the importance of early fault detection and diagnosis, which can help prevent equipment failure, reduce downtime, and increase efficiency [30].

3.3 Hybrid electric vehicle (HEV)

There are several new perspectives of Induction motors in HEVs including but not limited to improved fuel efficiency, reduced emissions, increased power density, improved energy recovery and advanced control techniques which are discussed below:

3.3.1 Improved fuel efficiency

Induction motors are known for their high efficiency, and they can improve the fuel efficiency of HEVs by providing additional power to the vehicle while the internal combustion engine (ICE) is operating at an optimal efficiency point. For example, a study by the National Renewable Energy Laboratory found that the use of induction motors in HEVs can improve fuel efficiency by up to 30% [12].

3.3.2 Reduced emissions

Induction motors produce no emissions, and they can be used to reduce the emissions of HEVs by providing power to the vehicle while the ICE is operating at an optimal efficiency point. For example, a study by the University of California, Berkeley found that the use of induction motors in HEVs can reduce emissions by up to 50% [16].

3.3.3 Increased power density

Induction motors can provide high power density, which means they can produce more power while taking up less space. This can be useful in HEVs, where space is often at a premium. For example, a study by the Electric Power Research Institute found that induction motors with high power density can be up to 50% smaller and lighter than traditional induction motors of the same power rating [24].

3.3.4 Improved energy recovery

Induction motors can be used in HEVs to improve the energy recovery during braking and deceleration, which can improve the overall energy efficiency of the vehicle. For example, a study by the Technical University of Munich found that the use of induction motors in HEVs can improve the energy recovery by up to 30% [27].

3.3.5 Advanced control techniques

Induction motors can be controlled using advanced techniques such as Model Predictive Control (MPC) and Artificial Intelligence (AI) techniques. These techniques can improve the performance and energy efficiency of induction motors in HEVs. For example, a study by the Technical University of Denmark found that MPC can improve the performance of induction motors in HEVs by up to 25% [28].

It is important to note that these studies are based on specific conditions, and further research is needed to confirm the results in other scenarios.

3.4 Wind turbines (WT)

Recent advances in the use of induction motors in wind turbine generators include the increased efficiency, increased power density, improved control, reduced harmonic distortion and increased fault-tolerance which are discussed below:

3.4.1 Increased efficiency

Induction motors have been shown to have high efficiency, which can improve the overall efficiency of the wind turbine generator. For example, a study by the National Renewable Energy Laboratory found that the use of induction motors in wind turbine generators can improve the overall efficiency by up to 96% [12].

3.4.2 Increased power density

Induction motors can provide high power density, which means they can produce more power while taking up less space. This can be useful in wind turbine generators, where space is often at a premium. For example, a study by the Electric Power Research Institute found that induction motors with high power density can be up to 50% smaller and lighter than traditional induction motors of the same power rating [24].

3.4.3 Improved control

Induction motors can be controlled using advanced techniques such as Direct Torque Control (DTC) and Field-Oriented Control (FOC), which can improve the speed and torque control of the wind turbine generator. For example, a study by the University of California, Berkeley found that DTC and FOC can improve the speed and torque control of induction motors in wind turbine generators by up to 50% [16].

3.4.4 Reduced harmonic distortion

Induction motors can be used to reduce harmonic distortion in wind turbine generators, which can lead to less noise and interference with other electrical equipment. For example, a study by the Technical University of Munich found that DTC can reduce harmonic distortion in induction motors in wind turbine generators by up to 50% [23].

3.4.5 Increased fault-tolerance

Induction motors can be designed to be fault-tolerant and able to maintain operation in the event of a fault, which can increase the overall reliability of wind turbine generators. For example, a study by the Technical University of Denmark found that induction motors with built-in fault-tolerance can improve the overall reliability of wind turbine generators by up to 40% [27].

4. Applications of induction motors

There are several applications of Induction motors on various fields including but not limited to pump systems and compressors which are discussed in this section:

4.1 Pump systems

Induction motors have been widely used in pump systems for many years, and new applications continue to be developed due to their efficiency, reliability, and low maintenance requirements. Some recent examples and common applications include VFDs, soft starting, adaptive control, water supply systems, Heat Ventilation Air Conditioning (HVAC) systems, Oil and gas industry, agriculture and Mining industry which are discussed below:

4.1.1 Variable frequency drive (VFD) control for energy efficiency

Induction motors used in conjunction with VFDs can improve the energy efficiency of pump systems by allowing the motor speed to be adjusted to match the system's demand. This can result in significant energy savings, as well as improved system performance and longer motor life [31].

4.1.2 Soft-starting for reducing inrush current

Induction motors used in pump systems can be designed with soft-starting features to reduce the inrush current and improve system performance. This can be accomplished through the use of reduced voltage starting methods, such as star-delta starting, or through the use of VFDs [32].

4.1.3 Adaptive control for improved system performance

Induction motors used in pump systems can be controlled using adaptive control methods to improve system performance and stability. This can be accomplished through the use of advanced control algorithms that adjust the motor speed and torque based on system conditions [33].

4.1.4 Water supply

Induction motors are used in various water supply systems such as municipal water supply, irrigation systems, and sewage treatment plants. They are used to power pumps that transfer water from one location to another [34–36].

4.1.5 HVAC systems

Induction motors are used in heating, ventilation, and air conditioning (HVAC) systems to power pumps that circulate hot or cold water through the system [34–36].

4.1.6 Oil and gas industry

Induction motors are used in the oil and gas industry to power pumps that move crude oil, natural gas, and other liquids through pipelines [34–36].

4.1.7 Agriculture

Induction motors are used in agricultural applications to power irrigation systems, and to pump water from wells or reservoirs [34–36].

4.1.8 Mining

Induction motors are used in mining applications to power pumps that remove water from underground mines and to transfer mining products from one location to another [34–36].

4.2 Compressors

The use of sensor data and machine learning algorithms to predict the failure of induction motors in compressor systems, allowing for preventative maintenance to be performed before a failure occurs [37]. Induction motors are widely used in compressors for various applications such as air conditioning, refrigeration, and natural gas processing. In recent years, there have been several advancements and developments in the application of induction motors in compressors. Some recent applications of induction motors in compressors are discussed below:

4.2.1 Variable frequency drive (VFD) controlled induction motors

The use of VFDs with induction motors has become increasingly popular in compressor applications. This technology enables the motor speed to be controlled, which leads to improved efficiency, reduced energy consumption, and lower maintenance costs [38].

4.2.2 High-efficiency induction motors

There has been a push towards improving the efficiency of induction motors in compressors. One way to achieve this is through the use of high-efficiency motors. These motors have higher efficiency levels and can significantly reduce energy consumption in compressor applications [39].

4.2.3 Permanent magnet synchronous motors (PMSMs)

PMSMs are another type of motor that has been gaining popularity in compressor applications. These motors have higher efficiency levels than induction motors and are ideal for applications that require high torque and low speed [40].

4.2.4 Sensor less control of induction motors

Recent developments in sensor less control technology have made it possible to eliminate the need for physical sensors in induction motor-driven compressors. This has the potential to reduce costs and increase reliability in compressor applications [41].

5. Conclusion

It is important to note that all the advances, new perspectives and applications of induction motors discussed in this chapter were based on the recent developments in the field of technology, science and innovation researched from the references mentioned in Section 6. There were several other advances and perspective approaches being carried out in the field of induction motors which needs further exploration and study. As technology continues to advance, we can expect to see more developments and improvements in this area.

In conclusion, induction motors have been a widely used technology for several decades due to their robustness, reliability, and efficiency. Recent advances in the field have focused on improving motor performance, reducing energy consumption, and enabling new applications. New perspectives on induction motor design have emerged, such as the use of new materials and innovative rotor designs. Additionally, the application of artificial intelligence and machine learning techniques has enabled new possibilities for predictive maintenance and fault diagnosis. Induction motors are widely used in many applications, including industrial machinery, transportation systems, and renewable energy. With ongoing research and development, induction motors are likely to continue to play a significant role in the industry for years to come.

Author details

Prabhu Thirugnanam
Engineering Office (Design Department), Heriot Watt University, Dubai, United Arab Emirates

*Address all correspondence to: talentedguy12@gmail.com

References

[1] Available from: https://www.nrel.gov/docs/fy13osti/55640.pdf

[2] Available from: https://www.energy.gov/eere/amo/downloads/variable-frequency-drives-technology-assessment

[3] Available from: https://www.sciencedirect.com/science/article/pii/S2095809917302761

[4] Available from: https://www.sciencedirect.com/science/article/pii/S0959652613005848

[5] Available from: https://www.sciencedirect.com/science/article/pii/S2405452618301865

[6] Energy Savings with Variable Frequency Drives. US Department of Energy. Available from: https://www.energy.gov/eere/articles/energy-savings-variable-frequency-drives

[7] High Efficiency Electric Motors. US Department of Energy. Available from:https://www.energy.gov/eere/articles/high-efficiency-electric-motors

[8] Permanent Magnet Synchronous Motors (PMSM) for Energy-Efficient Industrial Drive Systems. IEEE Transactions on Industry Applications. Available from:https://ieeexplore.ieee.org/document/8263078

[9] Available from: https://www.ijecce.org/article-details/Investigation-of-Energy-Saving-Methods-for-Induction-Motor-Driven-Systems-Using-VFDs-722.html

[10] Kaldellis JK, Karras KG. Energy efficiency of induction motors driven by variable frequency drives. Energy. 2010;**35**(11):4275-4281

[11] El-Refaie MA, Elsayed MA. Efficiency improvements in induction motors used in variable-frequency drives. IEEE Transactions on Industry Applications. 2010;**46**(2)

[12] Available from: https://www.nrel.gov/docs/fy16osti/65167.pdf

[13] Available from: https://www.j-eet.com/article/view/16051

[14] O'Neil DF, Smith RR. The use of variable frequency drives in process control applications. Control Engineering Practice. 2007;**15**(11):1419-1428

[15] Chiasson J. Advances in induction motor control. IEEE Industrial Electronics Magazine. 2011;**5**(3)

[16] Available from: https://www.eecs.berkeley.edu/Pubs/TechRpts/2005/EECS-2005-24.pdf

[17] Krishnan R. Sensorless control of induction motors: A survey. IEEE Transactions on Industrial Electronics. 2010;**57**(8)

[18] Guerrero JL. Intelligent control of induction motors: A review. IEEE Transactions on Industrial Electronics. 2010;**57**(10)

[19] Available from: https://iopscience.iop.org/article/10.1088/1742-6596/874/1/012070/meta

[20] Bose AK, Bhattacharya BK. Reliability and maintenance of induction motors driven by variable frequency drives. IEEE Transactions on Industrial Electronics. 2013;**60**(1):351-358

[21] Available from: https://www.medwelljournals.com/abstract/?doi=jeasci.2018.3244.3249

[22] Masoum MAS. A review of the impact of variable frequency drives on the reliability and maintenance of induction motors. Electric Power Components and Systems. 2014;**42**(4):354-367

[23] Available from: https://www.researchgate.net/publication/322177944_Reducing_harmonic_distortion_of_induction_motors_with_advanced_VFD_control_algorithms

[24] Available from:https://www.epri.com/abstracts/Pages/ProductAbstract.aspx?ProductId=000000003002006466

[25] Available from:https://ieeexplore.ieee.org/document/7726108

[26] Available from: https://www.sciencedirect.com/science/article/pii/S1877705815016033

[27] Available from: https://www.sciencedirect.com/science/article/pii/S095965261731711X

[28] Available from: https://www.sciencedirect.com/science/article/pii/S0360544219305737

[29] L. A. Aguirre, F. A. Silva, A. J. N. Lima, and R. S. Barbosa, Robust control for induction motor drives under parameter uncertainties" by in IEEE Transactions on Industrial Electronics, vol. 63, no. 12, pp. 7529-7540, 2016.

[30] Mirsalim M, Ehsani M. An intelligent condition monitoring framework for induction motors using deep learning techniques. IEEE Transactions on Industrial Informatics. 2019;**15**(11):5866-5875

[31] Stevens RL, Younkin GW. Energy savings with variable frequency drive control of induction motors. The Journal of Energy Engineering. 2007;**133**(4)

[32] Al-Haddad A, Al-Haddad B. Application of soft-starting in induction motors. The Journal of Engineering and Applied Sciences. 2009;**4**(8)

[33] Zarzoso MB, Yuz JI, Harley RG. Adaptive control of induction motor-driven pumps. IEEE Transactions on Control Systems Technology. 2017;**25**(5)

[34] ABB. Induction Motors for Pump Systems. 2019. Retrieved from: https://library.e.abb.com/public/126fe71d3b9888a7c12582b30060b898/Induction-motors-for-pump-systems-brochure-EN-2019.pdf

[35] ABB Motors and Generators. Applications of Induction Motors. 2020. Retrieved from: https://new.abb.com/motors-generators/industrial-motors/applications

[36] Siemens. Induction Motors for Pumps. 2021. Retrieved from https://www.siemens-energy.com/global/en/products-services/energy-efficient-motors/induction-motors-for-pumps.html

[37] Smith J, Patel A. Predictive maintenance of induction motors in compressor systems using machine learning. Journal of Predictive Maintenance. 2021;**15**(4)

[38] Yao X, Yang J, Guo J. Energy-saving control strategy of induction Motor for Centrifugal Compressor Based on improved particle swarm optimization. IEEE Access. 2019;7:91310-91318

[39] Gopalakrishnan M, Prakash R. Efficiency improvement of induction motor-driven compressor using an intelligent controller. International

Journal of Emerging Electric Power Systems. 2019;**20**(2):1-14

[40] Han L, Yang Y, Li J. Analysis of the application of permanent magnet synchronous Motor in Compressor. IEEE Access. 2020;**8**:118321-118331

[41] Huang Y, Xu J, Sun L. A Sensorless control strategy for induction motor-driven compressors based on fuzzy logic. IEEE Transactions on Industrial Electronics. 2019;**66**(6):4518-4527

Chapter 3

Application of Optimal Solutions in Field-Oriented Control of Induction Motor

Thinh Cong Tran, Pavel Brandstetter and Martin Kuchar

Abstract

This chapter presents the optimal solutions in field-oriented control for induction motors. It includes the introduction of the Genetic Algorithm (GA), Particle Swarm Optimizer (PSO), and Cuckoo Search Algorithm (CSA). In those algorithms, the GA algorithm is used to control the speed of the induction motor. With the GA algorithm, the speed response of the induction motor always achieves good stability in the case of no-load and under-load. This chapter also presents the optimal parameter search of the PID controller, such as proportional factor (*Kp*), integral coefficient (*Ki*), and differential coefficient (*Kd*), by the genetic algorithm. Finally, the simulation results on Matlab Simulink and the experimental results on TI's DSP 82335 have demonstrated the advantages of the proposed methods compared to the traditional PID methods.

Keywords: field-oriented control, optimal, induction motors, GA algorithm, CSA algorithm, PSO algorithm, PID, online, sensorless

1. Introduction

In today's industry, many induction motors are very much used. They account for approximately 50% of the energy in production activities. Speedy control of the induction motors is the main. The stable and accurate speed will improve product quality and work efficiency. The open-loop speed control will not guarantee accurate and stable speed [1, 2]. Some closed-loop speed control methods such as traditional PI and PID control, although better, cannot guarantee accurate and stable speed during operation, because with the value of K_p, K_i, and K_d if it works well at the beginning, usually in the middle or the end of the process, it is not necessarily good, or in cases where the load changes abnormally, for example, with load, no-load, large load, or small load [3–6]. In recent years, there have been smart solutions to improve this drawback, for example, using the ANN algorithm, PSO, fuzzy, etc. to find offline parameters K_p, K_i, and K_d for motor speed control systems [3–6, 7–12]. There are several methods of motor speed control, such as scalar control (V/Hz), torque control (torque), and vector control (FOC). Scalar control is an open-loop control of medium control quality, applied in systems that do not require high quality, such as fans and pumps. Torque control has the advantage of rapidly changing the torque of an induction motor. The two computational factors to control are the motor's torque and magnetic flux [13]. The vector-controlled IMs are extensively used in high-act motion controllers. Due to torque/flux separation, the vector controller attains good dynamic reaction and precise motioned controller like driving distinctly animated DC motors [14, 15]. Conversely, in a real-time operation, it is necessary to specify the exact motor parameters [16, 17]. To advance the quality of speed control of the motor, this chapter will present some intelligent algorithms, such as GA, PSO, and CSA algorithms, which will detail the application of GA algorithms in speedy control. Online-PID control of the induction motor with the FOC control model is selected to apply the GA algorithm in this chapter. The application of the GA algorithm allows the control program to find the optimal parameters K_p, K_i, and K_d during the control process in many situations, such as at no-load, under load, low load, or high load. We obtain a stable and accurate speed response during the control process.

2. Main topics in this chapter

2.1 Mathematical model of the FOC-controlled induction motor

The equations of the induction motor (IM) model in x-y coordinates are expressed as follows [1, 15]:

$$(R_s + pL_s)i_{sx} - L_s\omega_e i_{sy} + pL_m i_{rx} - L_m\omega_e i_{ry} = v_{sx} \tag{1}$$

$$L_s\omega_e i_{sx} + (R_s + pL_s)i_{sy} + L_m\omega_e i_{rx} + pL_m i_{ry} = v_{sy} \tag{2}$$

$$pL_m i_{sx} - L_m\omega_{sl} i_{sy} + (R_r + pL_r)i_{rx} - L_r\omega_{sl} i_{ry} = 0 \tag{3}$$

$$L_m\omega_{sl} i_{sx} + pL_m i_{sy} + (R_r + pL_r)i_{ry} + L_r\omega_{sl} i_{rx} = 0 \tag{4}$$

The electromagnetic torque in terms of x- and y-axes components is given by:

$$T_e = \frac{3}{2}P_p L_m (i_{sy}i_{rx} - i_{sx}i_{ry}) \tag{5}$$

Components of the rotor flux vector are described as follows:

$$\psi_{rx} = L_r i_{rx} + L_m i_{sx} \tag{6}$$

$$\psi_{ry} = L_r i_{ry} + L_m i_{sy} \tag{7}$$

From the above equations, after some transformations, we get the following expressions:

$$v_{sx} = R_s i_{sx} + \frac{L_s}{L_m}\frac{d\psi_{rx}}{dt} - L_m\sigma\frac{di_{rx}}{dt} - \omega_e\frac{L_m}{L_r}\psi_{ry} - \omega_e\sigma L_s i_{sy} \tag{8}$$

$$v_{sy} = R_s i_{sy} + L_s\sigma\frac{di_{sy}}{dt} + \omega_e L_s i_{sx} + \frac{L_m^2}{L_r}\frac{di_{sy}}{dt} + \omega_e L_m i_{rx} + L_m\frac{di_{ry}}{dt} \tag{9}$$

where $\sigma = 1 - \frac{L_m^2}{L_s L}$ is the leakage coefficient.

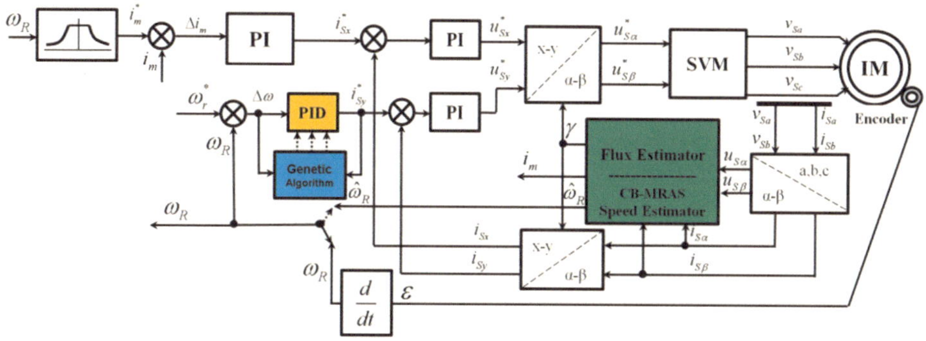

Figure 1.
The FOC speed control structure of the induction motor drive.

The rotor flux position θ_e is calculated from the rotor speed ω_r and slip frequency ω_{sl} as follows:

$$\theta_e = \int (\omega_r + \omega_{sl}) \mathrm{dt} \tag{10}$$

The current model of induction motor is given in following equation

$$i_{Sd} = i_{md} + T_R \cdot \frac{di_{md}}{dt} \tag{11}$$

$$i_{Sq} = i_{mq} + T_R \cdot \frac{di_{mq}}{dt} \tag{12}$$

Moreover, the magnitude and phase values of the magnetizing current are defined in equation

$$i_m = \sqrt{i_{m\alpha}^2 + i_{m\beta}^2} \tag{13}$$

$$\sin\gamma = \frac{i_{m\beta}}{i_m}, \quad \cos\gamma = \frac{i_{m\alpha}}{i_m} \tag{14}$$

From the above equation system, vector-controlled structure of induction motor is established as follows (**Figure 1**).

2.2 Traditional PID speed controller

The feature of the PID controller is the ability to use the three control terms of proportional, integral, and derivative influence on the controller output to apply precise control (see **Figure 2**).

From the above diagram, the classical PID speed controller for the induction motor drive can be expressed as follows:

$$i_{q\_ref} = K_p\left(\omega_{m\_ref} - \omega_m\right) + K_i \int \left(\omega_{m\_ref} - \omega_m\right) dt + K_d \frac{d\left(\omega_{m\_ref} - \omega_m\right)}{dt} \tag{15}$$

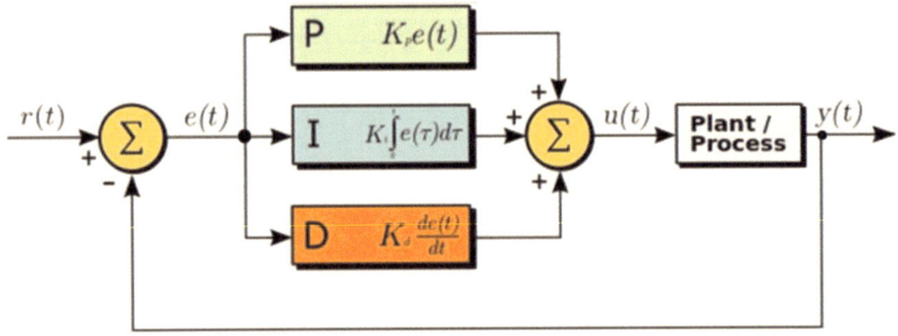

Figure 2.
The block diagram of the PID control.

The i_{q_ref} is the torque generating component called the reference current, the actual angular speed of the rotor is ω_m, the reference angular speed is ω_{m_ref}, K_p is the scaling factor, K_i is the integral factor, and K_d is the differential coefficient.

The actual value is compared with the reference value, and this deviation e_ω is handled by the PID speed controller. These parameters determine the characteristics of the electrical equipment [4, 5].

When starting the control cycle, normally the PID controller has a large scaling factor K_p to increase and decrease the speed quickly, the integral coefficient K_i must be small, and the differential coefficient K_d must be large to avoid overshoot. When the motor reaches the desired value, the scaling factor is small, the integral factor is large, and the differential factor is small to stabilize the motor speed at the reference value. The value of K_p is changed between K_{pmin} and K_{pmax}, K_i is changed between K_{imin} and K_{imax}, and K_d is changed between K_{dmin} and K_{dmax} to achieve satisfactory control performance.

The quality of the controller is evaluated according to the input parameters of the control loop as shown in **Figure 3**. The evaluation factors are t_r (the t_r is rise time), t_s (the t_s steady time), y_m (the y_m is overshoot), and Δ (the Δ is steady state error) [4].

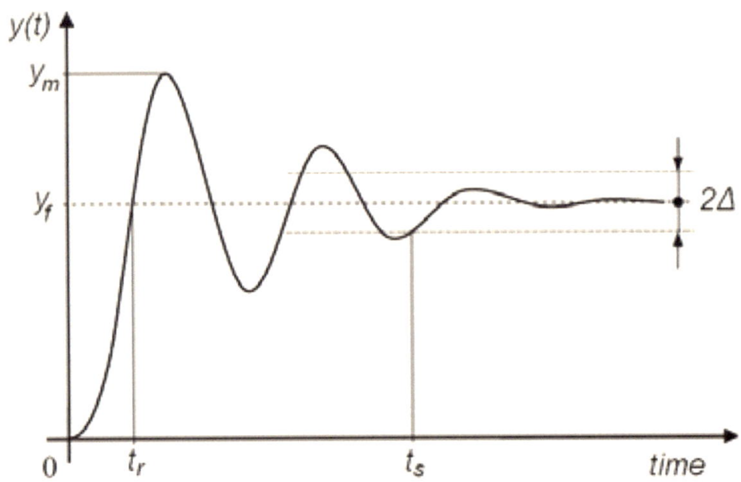

Figure 3.
The standard feedback for the control structure.

2.3 The intelligent algorithms

2.3.1 The genetic algorithm

This part will present the GA algorithm, how to determine cost function, a diagram of the GA-PID block in Matlab Simulink as well as details to implement algorithms for a speed-controlled structure of induction motor with a sensorless field-oriented model.

In this section, we will find the optimal parameters K_p, K_i, *and* K_d for PID controller by real number GA algorithm with the flowchart as follows (**Figure 4**):

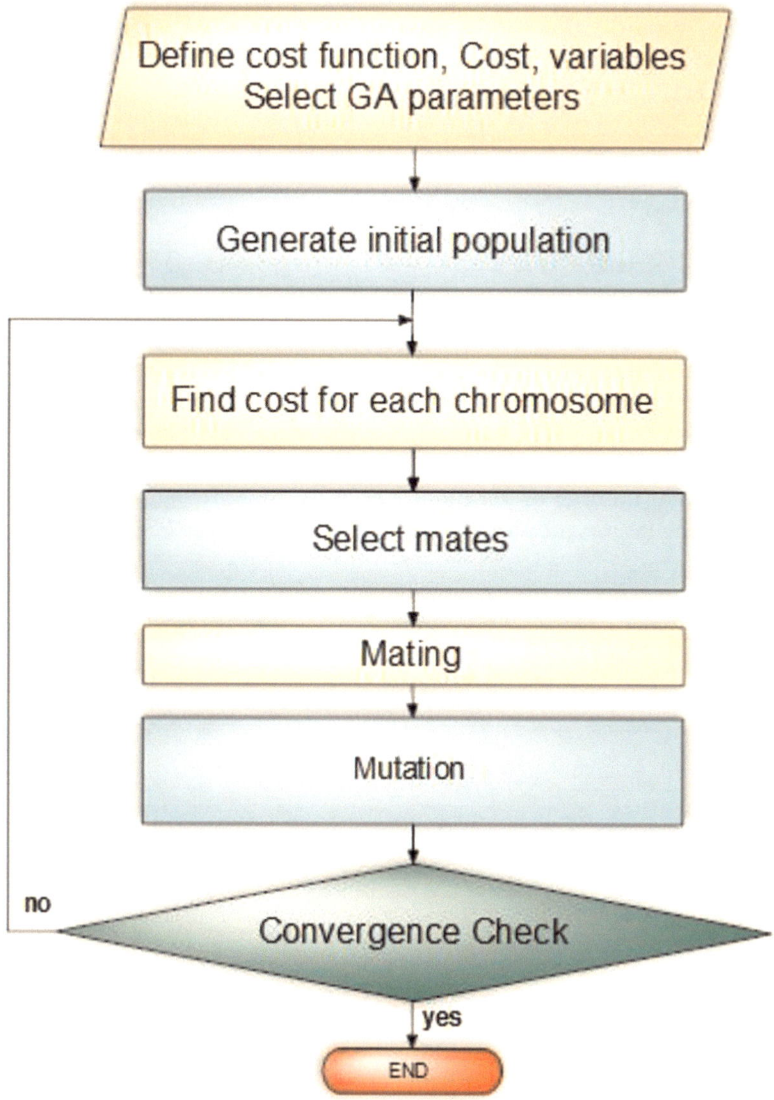

Figure 4.
The flowchart of the genetic algorithm.

2.3.2 The particle swam optimization (PSO) algorithm

The particle swam optimization algorithm is an optimization technique. It simulates the foraging of a school of fish or a flock of birds in nature [6, 7]. Simplicity, stable convergence, good computational efficiency, etc., are the advantages of this algorithm. Optimization problems often use this algorithm, such as adjusting the gain of the controller, recognizing parameters [3], and reducing losses in the power system [12]. With the PSO algorithm, the fish is considered a seed and forms a swarm. Each particle designates a membership outcome for the problem. Particles change their location by moving to the multidimensional search space. During the operation, each member changes its position according to the previous optimal position of each individual called P_{best}, and the overall optimal position called G_{best} as displayed in **Figure 5**.

In each step, the correction for the velocity and position of each individual is determined by using the current velocity and the space from *Pbest* to *Gbest*, as shown in Eqs. (16) and (17):

$$v_{ij} = wv_{ij} + c_1r_1\left(Pbest_{ij} - x_{ij}\right) + c_2r_2\left(Gbest - x_{ij}\right) \tag{16}$$

$$x_{ij} = v_{ij} + x_{ij} \tag{17}$$

$$\text{with} \quad w = w_{\max} - \frac{iter}{\max iter}(w_{\max} - w_{\min}) \tag{18}$$

Here, i: represents each individual, x_i: location of each ith individual, v_i: speed of ith individual, w: inaction function, $c_{1,2}$: acceleration coefficient (confident rate), and $\text{rand}_{1,2}$: a chance number over an interim [0,1], $Pbest_i$ is the best location created by the ith instance (best component), $Gbest_i$: best location created by flock (total best), last weightiness w_{max}, initial weightiness w_{min}, maximum number of iterations: *maxiter*, and current number of iterations: *iter*.

In **Figure 6**, the important phases in the progression of the PSO algorithm are shown below:

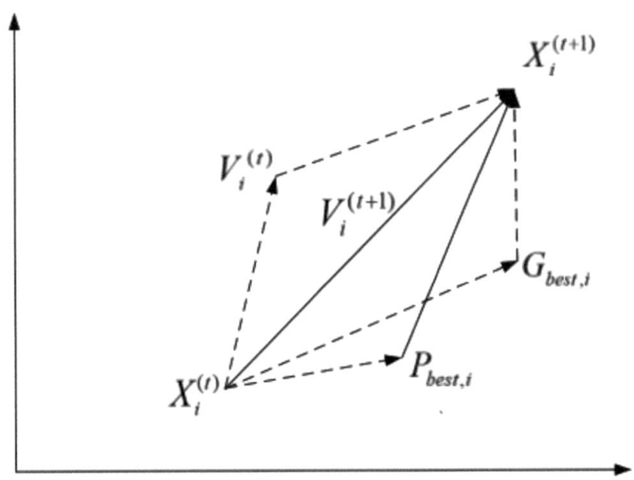

Figure 5.
Image of modifying velocity and particle location of the PSO process.

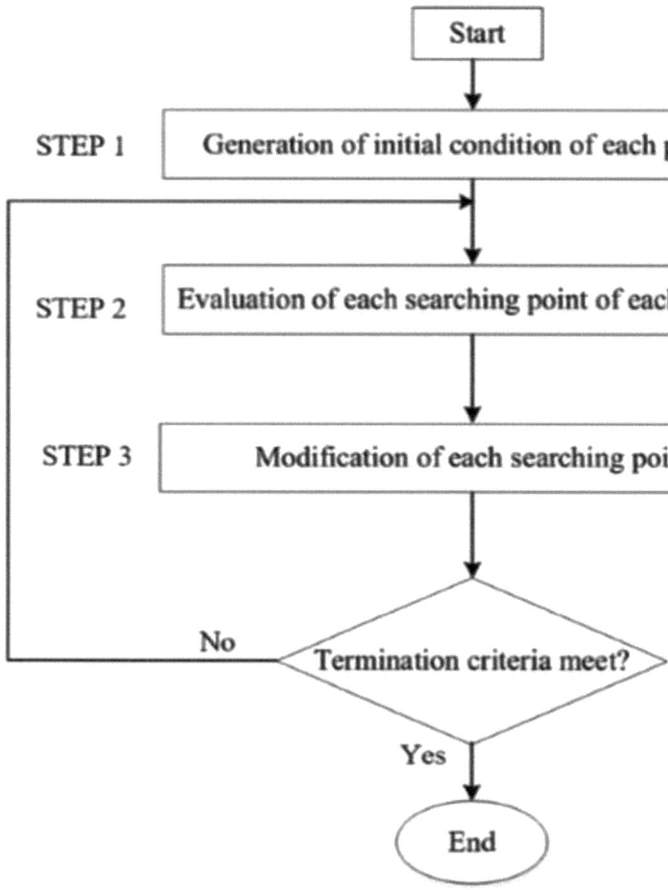

Figure 6.
Phases in the PSO procedure.

Step 1: Create a population of individuals whose position and speed are random in the d-dimensionality of the problematic universe within the allowed array.

Step 2: Calculate the cost function of each individual in the herd. If the present cost is better than the earlier *Pbest*, then *Pbest* is changed with the value of the present instance. If the previous value of *Gbest* is inferior to the best *Pbest*, then the value of *Pbest* will replace the value of *Gbest* and the position of the element with the most optimal value will be remembered.

Step 3: Transform the score found of each individual according to expression (17).

Step 4: Check the loop termination condition to control if it has been met (based on the cost function significance or the number of loops), if not, repeat the process from step 2. Else, the process is stopped.

2.3.3 The CSA algorithm (cuckoo search algorithm)

The key phases for CSA algorithm are shown below [8]:

1. **Initialization:** An inhabitant of host shells is represented by $X = [X_1, X_2, \dots, X_{Np}]^T$, here each shell $X_d = [P_{d1}, \dots, P_{dN}]; d = 1, \dots, Np$ expressive in place of the parameters, these parameters are initialized by:

$$P_{di} = p_{i\min} + rand_1{}^* (p_{i\max} - p_{i\min});\ i = 1, \dots, N \tag{19}$$

   where rand is a distributed random number in [0, 1] for each population of host nests. The maximum and minimum of the parameters are defined as the P_{imin} and P_{imax}.

   The establishment residents of the host shells have the greatest significance of each shell $Xbest_d$ $(d = 1, \dots, N_d)$ and the shell matches the greatest objective function and has the greatest shell *Gbest* among all the shells in resident.

2. **Creation of new result by Lévy flights:** The novel result of each shell is produced below:

$$X_d^{new} = Xbest_d + \alpha^* rand_2{}^* \Delta X_d^{new} \tag{20}$$

   The objective function will be reexamined to find the new greatest value of each shell $Xbest_d$ and the best of all shells *Gbest* by linking the stored calculated value and the new calculated value created.

3. Strange egg detection and randomization

   A strange egg in the shell of a host bird with probability P_a is detected. A new result similar to Lévy flights generated by this event is shown below:

$$X_d^{dis} = Xbest_d + K^* \Delta X_d^{dis} \tag{21}$$

   Here, K is the coefficient updated and determined based on the probability that the host bird detects an alien egg in their shell.

$$\begin{aligned} &\textit{if } rand_3 < p_a \textit{ then } K = 1 \textit{ else } K = 0 \\ &\Delta X_d^{dis} = rand_4{}^* \left(randp_1(Xbest_d) - randp_3(Xbest_d)\right) \end{aligned} \tag{22}$$

   Like the result of Lévy flights, this novel result is also resolute again for each shell $Xbest_d$ and the greatest importance of all shells *Gbest* is established based on significance acquired from (21).

4. **Stopping conditions:** The CSA algorithm is finished when the recent repetition is like or larger than the maximum number of repetitions (**Figure** 7).

2.4 The speed controller of IM using genetic algorithm

To perform a genetic algorithm as well as some other soft computing algorithms, it is very important to find the cost function correctly, below is how to define it.

From equation in the continuous time:

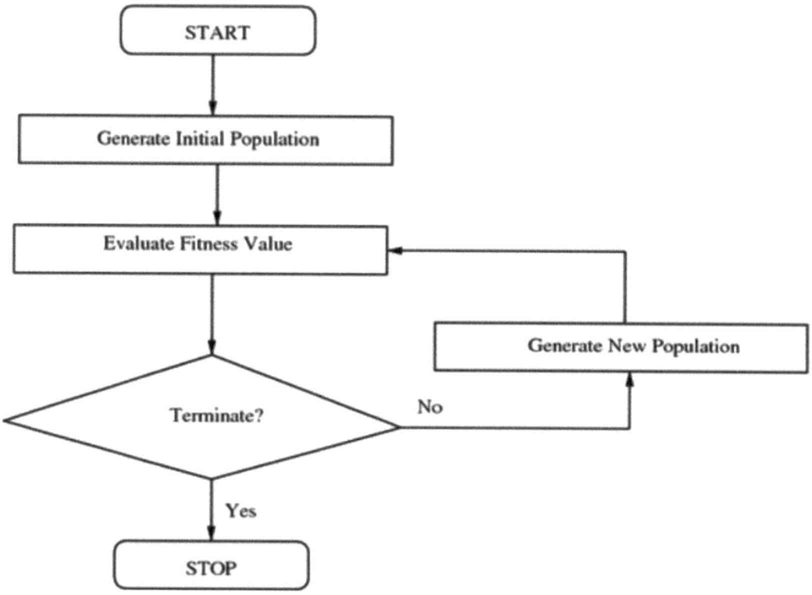

Figure 7.
Steps in CSA algorithm.

$$u(t) = K_p e(t) + K_i \int_0^t e(t)dt + K_d e(t) \tag{23}$$

To determine the objective function, the moment-generating component is represented in the discrete form with the sampling cycle ΔT as shown below:

$$\int_0^{t_k} e(\tau)d\tau = \sum_{i=1}^{k} e(t_i)\Delta t \tag{24}$$

and

$$\frac{de(t_k)}{dt} = \frac{e(t_k) - e(t_{k-1})}{\Delta t} \tag{25}$$

Here, the endless time inaccuracy at the ith sampling time is $e(t_i)$. Then, the expression becomes:

$$u(t_k) = K_p \left(e(t_k) + \frac{\Delta t}{T_i} \sum_{i=1}^{k} e(t_i) + \frac{T_d}{\Delta t} (\mathrm{e}(\mathrm{t}_k) - \mathrm{e}(\mathrm{t}_{k-1})) \right. \tag{26}$$

The $u(t_k)$ can also be calculated from $u(t_{k\text{-}1})$ as follows:

$$u(t_k) = u(t_{k-1}) + K_p \left(1 + \frac{\Delta t}{T_i} + \frac{T_d}{\Delta t}\right) \left(e(t_k) + K_p \left(-1 - \frac{2T_d}{\Delta t} \right) \mathrm{e}(\mathrm{t}_{k-1}) + K_p \frac{T_d}{\Delta t} e(t_{k-2}) \right. \tag{27}$$

$$\text{with}: \quad K_i = T_i/K_p; K_d = K_p T_d$$

$$e_k = \omega_{m\_ref(k)} - \omega_{m(k)}$$

$$u(t_k) - u(t_{k-1}) = \Delta_{sy\_ref} = i_{sy\_ref(k)} - i_{sy\_ref(k-1)}$$

Eq. (27) is rewritten as shown below:

$$i_{sy\_ref\,k} = i_{sy\_ref\,k-1} + K_p\left(1 + \frac{\Delta t}{T_i} + \frac{T_d}{\Delta t}\right)e_k + K_p\left(-1 - \frac{2T_d}{\Delta t}\right)e_{k-1} + K_p\frac{T_d}{\Delta t}e_{k-2} \quad (28)$$

2.5 The detail block and the steps to implement GA-PID speed controller

From this expression, the designed schematic of GA-PID block is as below (**Figure 8**).

The right-hand side of expression (28) will be evaluated to collect the new values of the K_p, K_i, and K_d. All parameters in the overhead equation, except $i_{sy\_ref\,k}$, are known at the $(k-1)^{\text{th}}$ sampling point. If the different values of K_p, K_i, and K_d are chosen, then obviously the different reactions of the plant will be collected. Therefore, the parameter adjustment problem of the PID controller can be considered by choosing the three parameters K_p, K_i, and K_d, so that the response of the plant will be as desired to automatically detect PID parameter online.

2.6 Simulation results

The Matlab Simulink is used to simulate the controller of an induction motor with the following parameters: P = 1HP, U_{dc} = 300 V, P_p = 2, R_S = 2.1 Ω, R_r = 1.51 Ω,

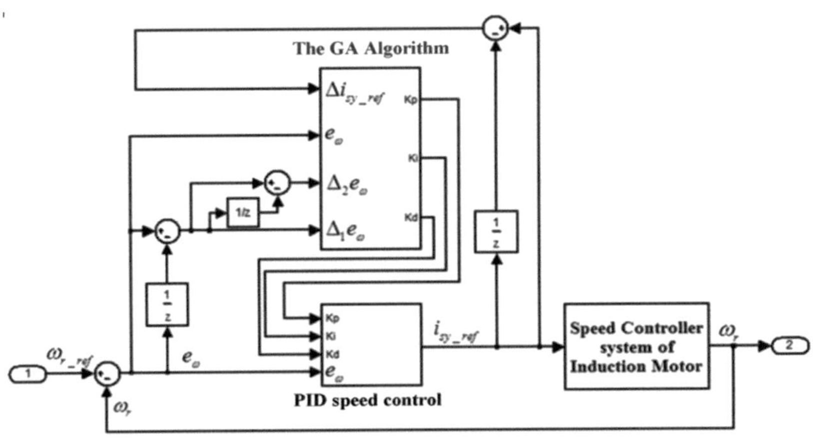

Figure 8.
The speed controller using genetic algorithm.

Parameter	GA speed controller	Classical PI controller
Proportionate coefficient K_p	[0–50] A/rad/s	90 A/rad/s
Integral coefficient K_i	[0–50] A/rad	90 A/rad
Derivative coefficient K_d	[0–50] A/rad.s^2	2A/rad.s^2
Generations number (i)	3	
The coefficients K_p, $K_{i,}$ and K_d in populace (j)	8	

Table 1.
The speedy control has the based parameters as below.

L_m = 0.129H, J = 0.043 Kg.m^2, and L_S = 0.137 H. The list of control parameters is given in **Table 1**.

Figures 9–16 display important characteristics of speed controller of the induction motor under different operating conditions.

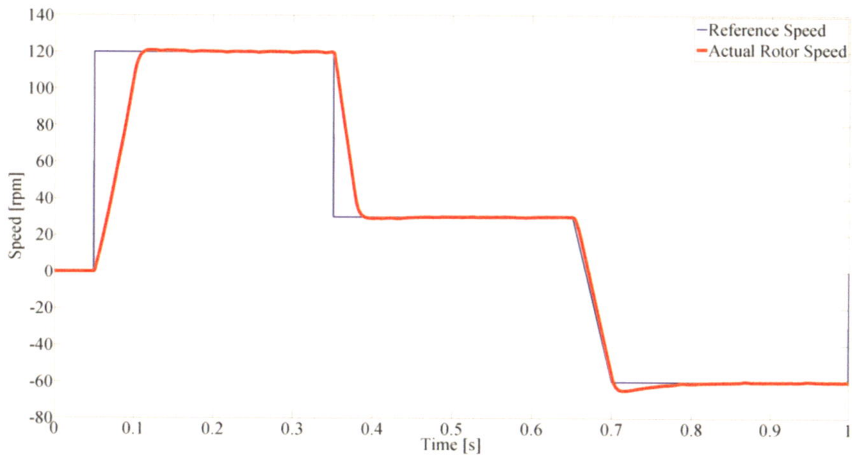

Figure 9.
Reference speed ω_{m_ref} (blue) and actual rotor speed ω_m (red) of the IM with traditional PID speed controller without load.

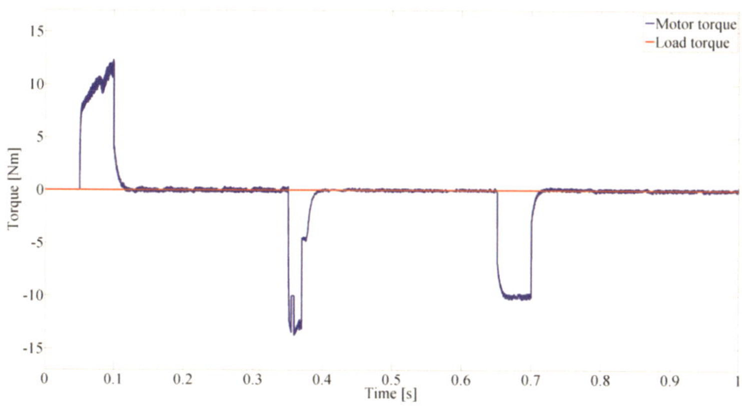

Figure 10.
Load torque (red) and induction motor torque (blue) with traditional PID speed controller.

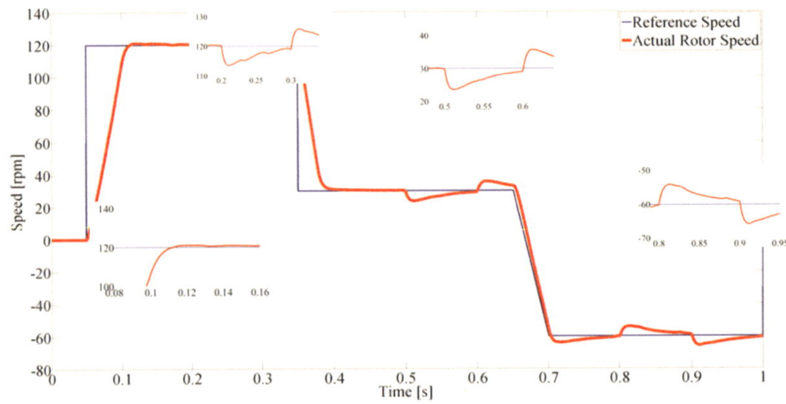

Figure 11.
Reference speed ω_{m_ref} (blue) and actual rotor speed ω_m (red) of the IM with traditional PID speed controller has the load.

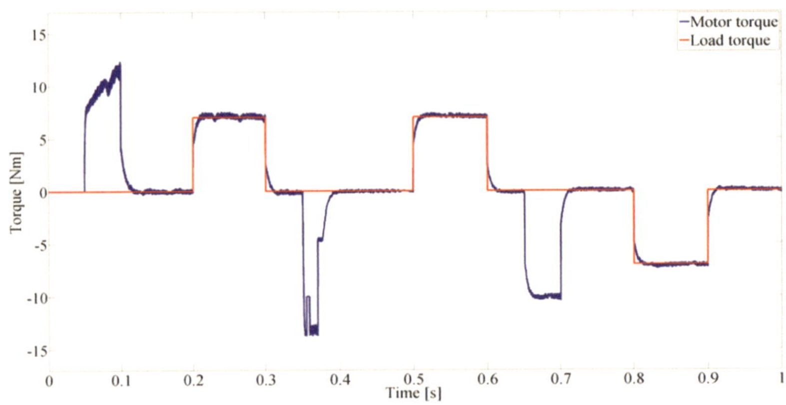

Figure 12.
Load torque (red) and induction motor torque (blue) with traditional PID speed controller.

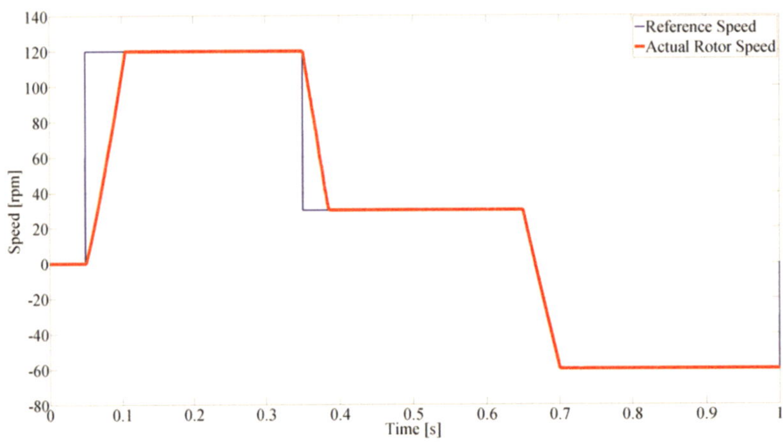

Figure 13.
Reference speed ω_{m_ref} (blue) and actual rotor speed ω_m (red) of the IM with the online PID speedy control using the GA algorithm without load.

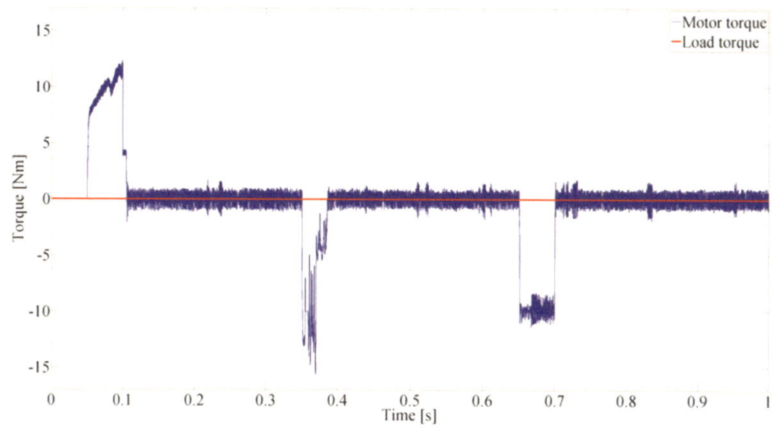

Figure 14.
Load torque (red) and induction motor torque (blue) with the online PID speedy control using the GA algorithm.

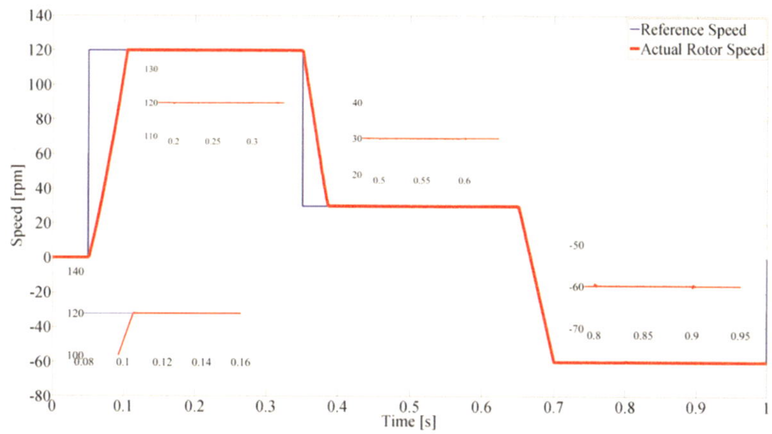

Figure 15.
Reference speed ω_{m_ref} *(blue) and actual rotor speed* ω_m *(red) of the IM with the online PID speedy control using the GA algorithm has the load.*

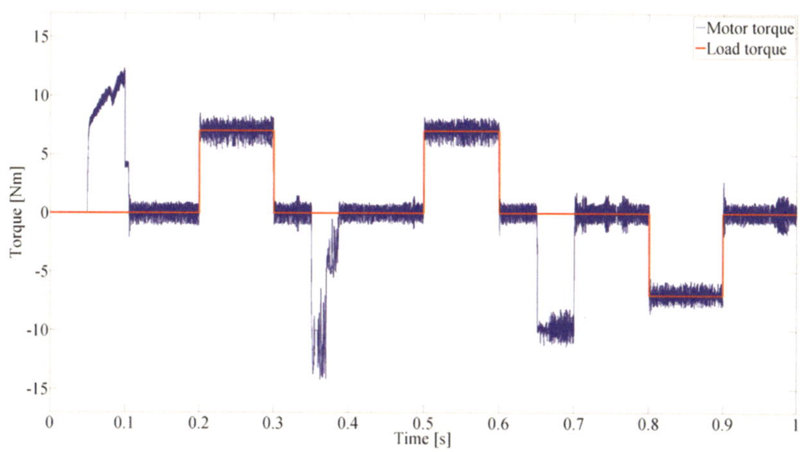

Figure 16.
Load torque (red) and induction motor torque (blue) with the online PID speedy control using the GA algorithm.

2.6.1 Traditional PID speed controller

2.6.1.1 Traditional PID speed controller without load

Under the no-load condition, the speed response of the PID controller for the induction motor is quite good (see **Figure 9**).

Figure 10 shows the load torque and induction motor torque.

2.6.1.2 Traditional PID speed controller with load

For a traditional PID speedy control in case of load, the speed response varies quite a lot.

2.6.2 The online PID speedy control using the GA algorithm

2.6.2.1 The online PID speedy control using the GA without load

With the PID speedy control using the GA algorithm to optimize the parameters K_p, K_i, *and* K_d, in the case of no load or load, the speed response is still stable (see **Figures 13** and **15**).

2.7 Experimental results

This experiment was performed in the laboratory. Our experimental model is shown below, the equipment includes the TMS 320F82335 board, the inverter with the Semikron IGBT, the drive supply modulated voltage for the induction motor, 3-phase induction motor, the generator load for a motor, 3-phase transformer output voltage

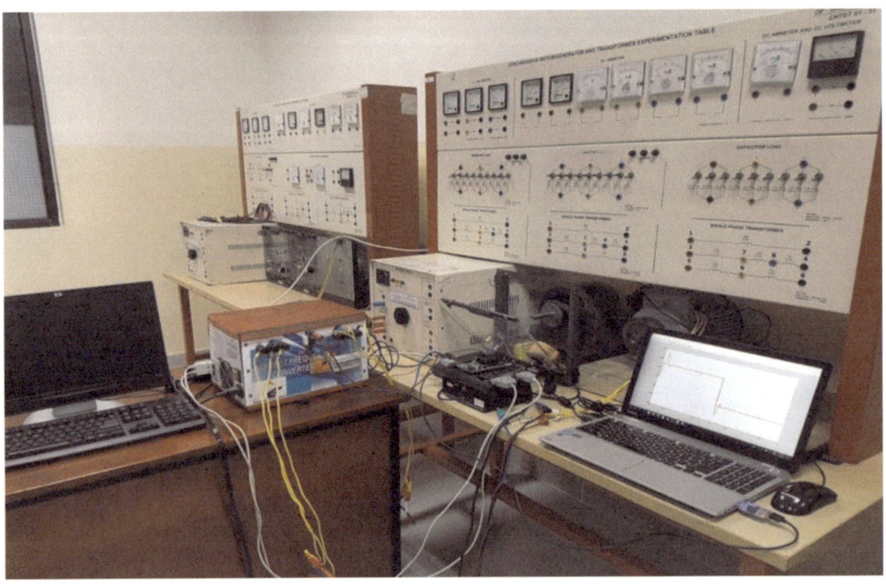

Figure 17.
The DSP-28335 board and IM motor used in experiment.

varies from 0–220 Vac. We did experiment for both control methods above such as the simulation part at different speed in two cases no-load and load. The figure below shows the results (**Figure 17**).

2.7.1 The traditional PID speed control without load

In actual performance, the traditional PID speed controller has a rather large overshoot compared to the PID using the genetic algorithm (see **Figure 18**).

2.7.2 The online GA-PID speed control without load

In the experiment with the PID speed controller using a genetic algorithm to optimize control parameters, the speed response is still stable in the no-load case (see **Figure 19**).

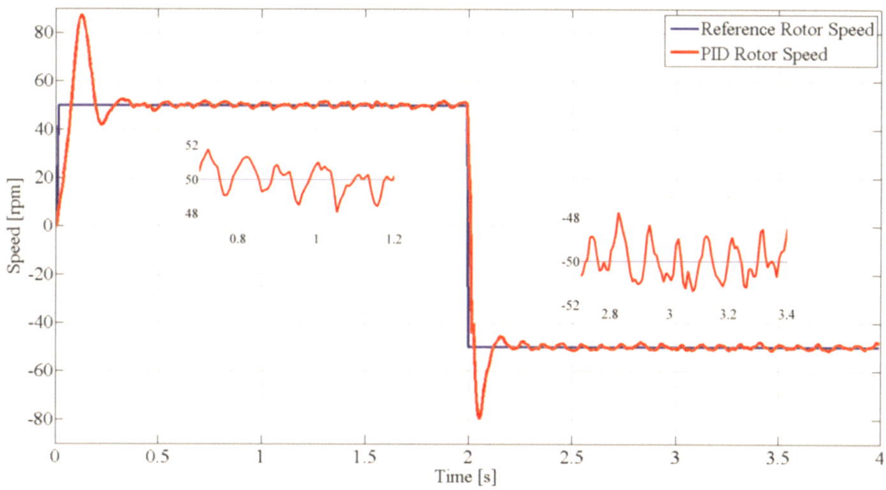

Figure 18.
Real rotor speed ω_m (red) and mention speed ω_{m_ref} (blue) of the induction motor.

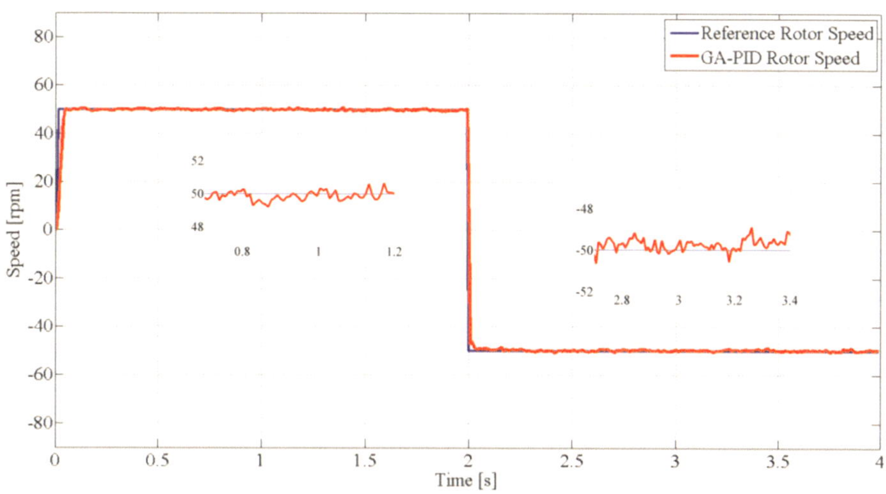

Figure 19.
Reference speed ω_{m_ref} (blue) and actual rotor speed ω_m (red) of the IM in the online GA-PID speed control without load.

Compare the speed response of the two methods on the same coordinate system without load (see **Figure 20**).

2.7.3 The traditional PID speed control with load

Figures 21 and **22** show the speed and torque response of the IM according to the traditional PID control model. The speed response of an induction motor applying the online genetic algorithm remains stable under load (see **Figure 23**).

Figure 24 depicts the load torque and IM torque in the online GA-PID speed control with the load. The comparison of the speed response with the load of the IM between the traditional PID controller and the online GA-PID controller is shown in **Figure 25**.

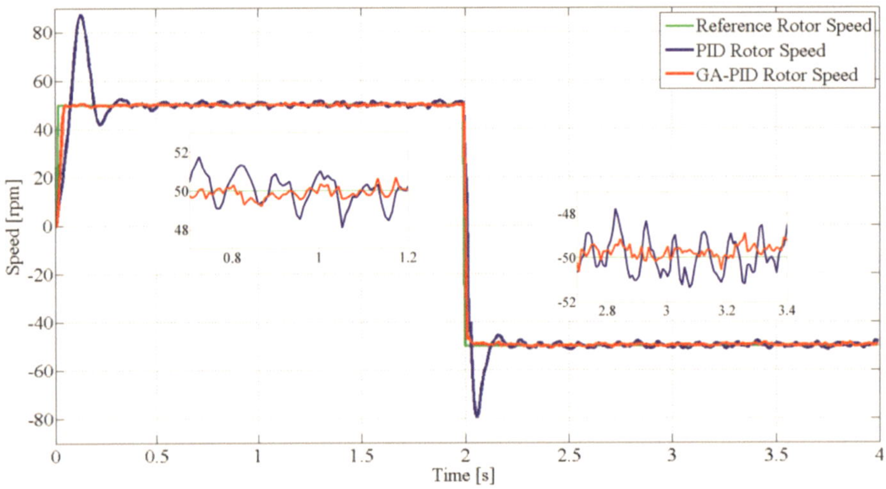

Figure 20.
Real rotor speed ω_m (red) and desired speed ω_{m_ref} (blue) of the IM in the above two methods without load.

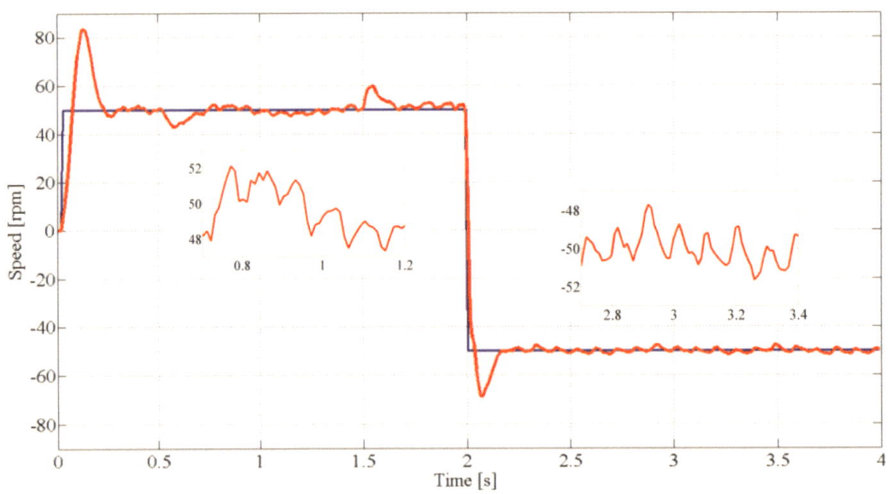

Figure 21.
Reference speed ω_{m_ref} (blue) and actual rotor speed ω_m (red) of the IM in the traditional PID speed control with load.

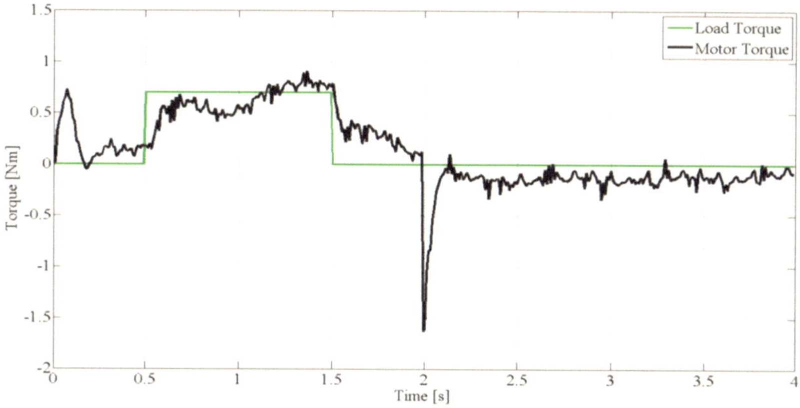

Figure 22.
The load torque (green) and IM torque (black) of the PID speed controller with load.

2.7.4 The online GA-PID speed control with load

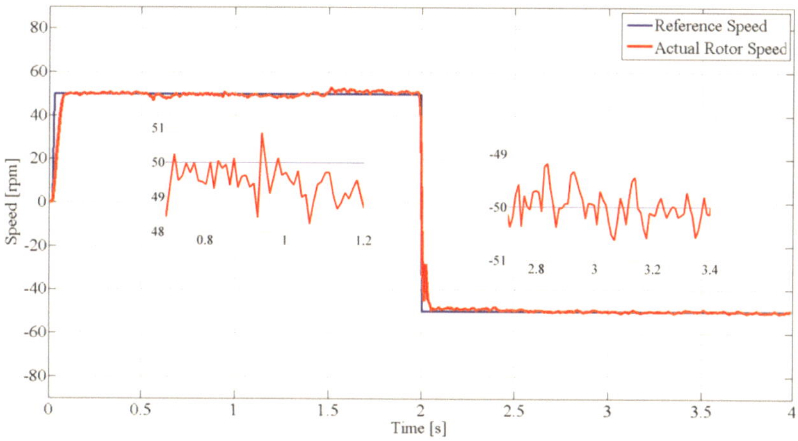

Figure 23.
Reference speed ω_{m_ref} (blue) and actual rotor speed ω_m (red) of the IM in the online GA-PID speed control with load.

3. Conclusion

This chapter presented some intelligent algorithms such as GA, PSO, and CSA algorithms to apply to the control model of induction motors. In the paper, the application of the GA algorithm in the speed control of induction motors has been presented in detail. The speed control of the induction motor by conventional PID in case of no-load or unchanged working conditions is quite good, but under variable conditions, the control quality is not guaranteed. In the case of speed control of induction motor using PID controller where the optimal proportional parameter K_p, integral K_i, and differential K_d are updated continuously by genetic algorithm, the speed response is very good and stable under load or no-load conditions. These have been proven through simulation results on Matlab Simulink as well as in experiment with TI's DSP-28335 board above.

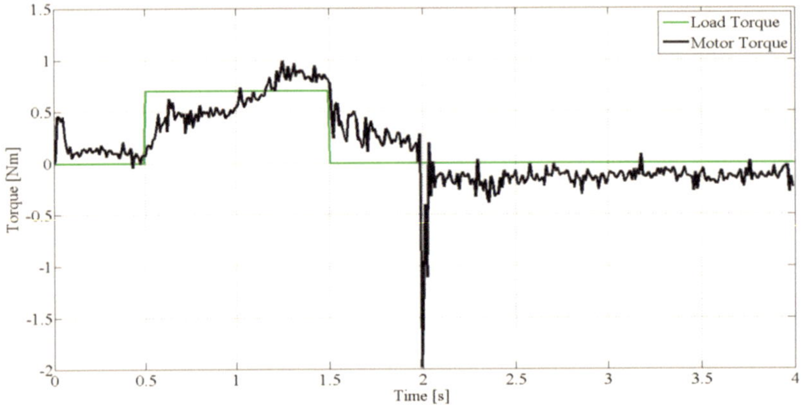

Figure 24.
Load torque (green) and IM torque (black) of in the online GA-PID speed control with load.

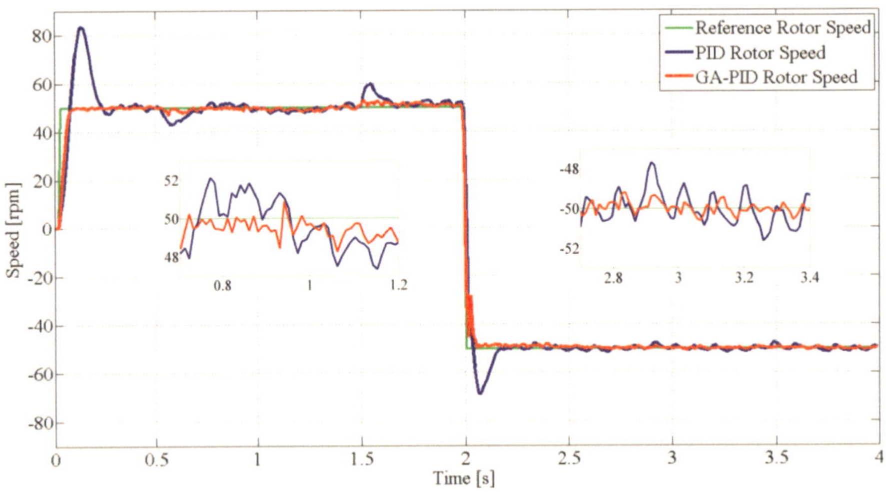

Figure 25.
Reference speed ω_{m_ref} (green), PID rotor speed of the IM ω_{est} (blue), and GA-PID rotor speed ω_m (red) of the IM with load.

In the content of this topic, having mentioned the speed control for induction motors by field-oriented method, readers can apply the algorithms mentioned in the torque control model for IM motors or search motor parameters such as stator resistance, rotor, stator winding inductance, rotor, and rotor time constant (because the quality of IM motor control according to torque model or field orientation depends heavily on the parameters of IM motor).

In addition, other motors such as PMSM motors and Brushless DC motors are also widely applied in the field of industrial robots, machines with high precision control needs, due to strong torque and higher working efficiency than induction motor. Therefore, the application of the presented algorithms: GA, PSO, CSA algorithm, etc., to the optimal control of these motors (PMSM motor, Brushless DC motor, etc.) as well as develop more algorithms and other smarts such as Artificial Neural Network

(ANN), Fuzzy, sliding mode control, extended Kalman filter (EKF), or a combination of these algorithms will contribute to bringing good efficiency to our industry and lives.

Author details

Thinh Cong Tran[1*], Pavel Brandstetter[2] and Martin Kuchar[2]

1 Faculty of Electrical and Electronics Engineering, Ton Duc Thang University, Vietnam

2 Faculty of Electrical Engineering and Computer Science, VSB–Technical University of Ostrava, Czech Republic

*Address all correspondence to: trancongthinh@tdtu.edu.vn

References

[1] Vas P. Vector Control of AC Machines. Oxford: Clarendon Press; 1990

[2] Vas P. Artificial-intelligence-based electrical machines and drives: application of fuzzy, neural, fuzzy-neural, and genetic-algorithm-based techniques. Oxford University Press; 1999

[3] Banerjee T, Sumana C, Jitendranath B, Abhisek M, et al. Off-line optimization of PI and PID controller for a vector controlled induction motor drive using PSO. In: Proceedings of International Conference on Electrical & Computer Engineering (ICECE). IEEE. 2010. pp. 74-77. DOI: 10.1109/ICELCE.2010.5700556

[4] Tran TC, Brandstetter P, Duy VH, Vo HH, Dong C. PID speed controller optimization using online genetic algorithm for induction motor drive. In: International Conference on Advanced Engineering Theory and Applications. Cham: Springer; 2016. pp. 564-576. DOI: 10.1007/978-3-319-50904-4_60

[5] Laroussi K, Zelmat M, Rouff M. Implementation of a fuzzy logic system to tune a PI controller applied to an induction motor. Advances in Electrical and Computer Engineering. 2009;**9**(3): 107-113. DOI: 10.4316/AECE.2009.03019

[6] Rekha M, Kumar MK. Variable frequency drive optimization using torque ripple control and self-tuning PI controller with PSO. International Journal of Electrical and Computer Engineering (IJECE). 2019;**9**(2):802-814. DOI: 10.11591/ijece.v9i2.pp802-814

[7] Dang HS, Palacky P, Kuchar M, Brandstetter P, Tran CD. Particle swarm optimization-based stator resistance observer for speed sensorless induction motor drive. International Journal of Electrical and Computer Engineering. 2021;**11**(1):815. DOI: 10.11591/ijece.v11i1.pp. 815-826

[8] Hassan AY, El-latifBadr MA, Wahsh SA E-m. Cuckoo search based real time implementation of direct torque control of PMSM. In the 2018 Twentieth International Middle East Power Systems Conference (MEPCON). IEEE; 2018. pp. 235-241. DOI: 10.1109/MEPCON.2018.8635189

[9] Kennedy J, Eberhart R. Particle swarm optimization. In: Proceedings of ICNN'95-international conference on neural networks; 1995. IEEE. vol. 4, pp. 1942-1948. DOI: 10.1109/57ICNN.1995.488968

[10] Shi Y, Eberhart R. A modified particle swarm optimizer. In: Proceedings of the IEEE International Conference on Evolutionary Computation. IEEE. 1998. pp. 69-73. DOI: 10.1109/ICEC.1998.699146

[11] Karanayil B et al. Online stator and rotor resistance estimation scheme using artificial neural networks for vector controlled speed sensorless induction motor drive. IEEE Transactions on Industrial Electronics. 2007;**54**(1): 167-176. DOI: 10.1109/TIE.2006.888778. DOI: 10.1109/ICELCE.2010.5700556

[12] Tofighi EM, Mahdizadeh A, Feyzi MR. Online estimation of induction motor parameters using a modified particle swarm optimization technique. In: Proceedings of IECON 2013-39th Annual Conference of the IEEE Industrial Electronics Society. IEEE. 2013. pp. 3645-3650. DOI: 10.1109/IECON.2013.6699715

[13] Mini R, Shabana Backer P, Satheesh H, M. N. Dinesh: Low speed estimation of sensorless DTC induction motor drive using MRAS with neuro fuzzy adaptive controller. International Journal of Electrical and Computer Engineering (IJECE). 2018;**8**(5): 2691-2702. DOI: 10.11591/ijece.v8i5. pp2691-2702

[14] Sreejeth M, Singh M, Kumar P. Efficiency enhancement for indirect vector-controlled induction motor drive. International Journal of Electronics. 2019;**106**(9):1281-1294. DOI: doi.org/ 10.1080/00207217.2019.1584921

[15] Girovsky P, Timko J, Zilkova J. Shaft sensor-less FOC control of an induction motor using neural estimators. Acta Polytechnica Hungarica. 2012;**9**(4):31-45

[16] Karanayil B et al. Stator and rotor resistance observers for induction motor drive using fuzzy logic and artificial neural networks. IEEE Transactions on Energy Conversion. 2005;**20**(4):771-780. DOI: 10.1109/TEC.2005.853761

[17] Vasic V et al. A stator resistance estimation scheme for speed sensorless rotor flux oriented induction motor drives. IEEE Transactions on Energy Conversion. 2003;**18**(4):476-483. DOI: 10.1109/TEC.2003.816595

Chapter 4

Association Induction Motors to Converters

Fezazi Omar and Ayad Ahmed Nour el Islam

Abstract

Induction motors feature state-of-the-art technology ensures high reliability, energy efficiency, versatility, and long life. They are also called "squirrel cage" or "induction" motors and can be single or three-phase. Because it is simple to use and install, induction motors are the most used motors. It has a long life because it is durable and wear-free. It does not need a lot of maintenance (maintenance costs for an induction motor are very low compared to other types of electric motors). It can be modified for a variety of applications, such as changing the motor's rotational speed using a frequency converter. It is a low-cost motor. One example of a use for these motors is the industrial industry (mostly the industry of machines-tools) and the appliances (the wash-cloths). The creation of cold (via condensers and refrigerators). The pumping (pool pumps, lifting pumps, etc.). Ventilo-convecteurs or ventilation air restrained (compressors). In this chapter, the induction motor is trained and controlled by different converters, and Matlab SimPowerSystem is used to simulate the different methods used.

Keywords: association, induction motors, converters, electrical control, Matlab SimPowerSystem

1. Introduction

An asynchronous motor, also known as an induction motor, is a type of electric motor that operates without the need for external electrical connections to the rotor. The motor works on the principle of electromagnetic induction [1]. In an asynchronous motor, the stator, which is the stationary part of the motor, generates a rotating magnetic field that induces a current in the rotor, the rotating part of the motor. As a result, a magnetic field is created in the rotor, which causes it to rotate, thus driving the mechanical load [2–4].

Asynchronous motors are commonly used in industrial and commercial applications due to their robustness, reliability, and low cost. They are also known for their simple design, low maintenance requirements, and ability to operate under varying load conditions [2–4].

There are two types of asynchronous motors: the squirrel cage motor and the wound rotor motor. In the squirrel cage motor, the rotor consists of a cylindrical core made of laminated steel with conductive bars or "squirrel cages" embedded in the

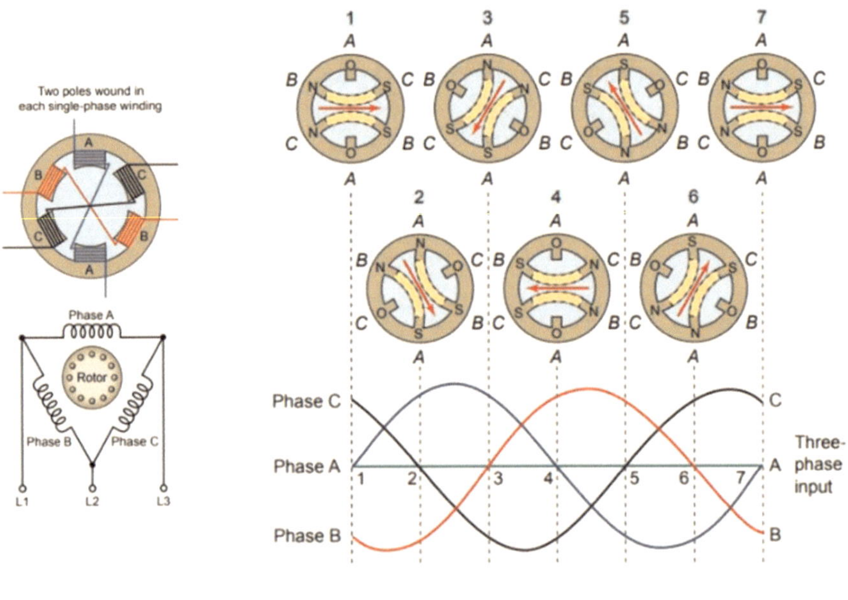

Figure 1.
Rotating magnetic field [1].

surface. The wound rotor motor has a rotor with a series of windings connected to slip rings, allowing external resistors to be connected for speed control [5, 6].

Asynchronous motors can operate on a wide range of voltages and frequencies and are commonly used in industrial applications such as pumps, fans, compressors, and conveyors. **Figure 1** illustrates the concept of a rotating magnetic field as it applies to the stator of a three-phase induction motor.

2. Using SimPowerSystems in control systems

One of the key advantages of using SimPowerSystems in the design and simulation of control systems for asynchronous motors is that it can help us to approach reality more closely than simple theoretical models [7, 8].

SimPowerSystems allows engineers and researchers to create virtual models of electrical power systems, including asynchronous motors and their associated control systems. These models can be used to simulate the behavior of the system under different operating conditions, such as varying loads and speeds [7, 8].

SimPowerSystems uses advanced mathematical models to represent the electrical, magnetic, and mechanical behavior of the system. These models take into account real-world phenomena such as magnetic saturation, rotor resistance, and voltage drops in the stator and rotor circuits. By using these detailed models, SimPowerSystems can provide a more accurate representation of the system's behavior than simpler theoretical models.

SimPowerSystems also supports hardware-in-the-loop (HIL) simulation, which allows engineers to test their control systems using physical hardware, such as motor drives and controllers. This can help to further improve the accuracy of the simulation and identify any issues or limitations in the control system before implementing it in a real-world application [7, 8].

SimPowerSystems can help engineers to design and optimize control systems for asynchronous motors that are more accurate and realistic than simple theoretical models. This can help to improve the performance and reliability of the system while reducing costs and development time.

3. Asynchronous motor in Matlab simscape (SimPowerSystem)

To model an asynchronous motor using SimPowerSystems in SI units, a new Simulink model must be created. Once the model is created, engineers can then choose the appropriate asynchronous motor block from the SimPowerSystems library and set its parameters, such as the rated voltage, rated frequency, and the number of poles. The asynchrounous machine SI unites is shown in **Figures 2** and **3**.

Change the wound asynchronous motor to squirrel-cage asynchronous motor and choose one motor from *squirrel-cage preset mode.*

3.1 Rated parameter three-phase asynchronous motor

When driving a three-phase asynchronous motor underrated (nominal) parameters, it is important to ensure that the motor is operated within its designed voltage, current, and frequency range. This includes [2–6]:

- Voltage: The motor should be supplied with the nominal voltage specified by the manufacturer. Operating the motor at a voltage higher or lower than its nominal voltage can cause overheating, reduced efficiency, and premature failure.

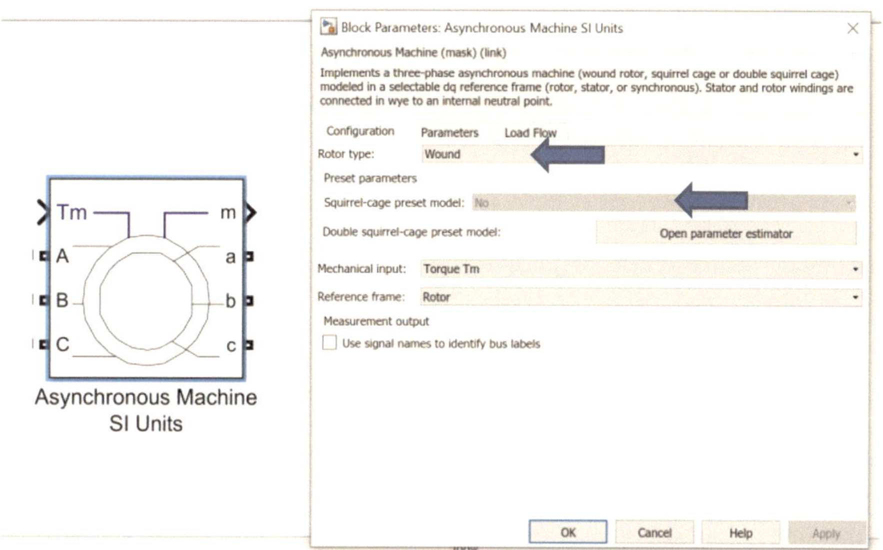

Figure 2.
Wound asynchronous motor.

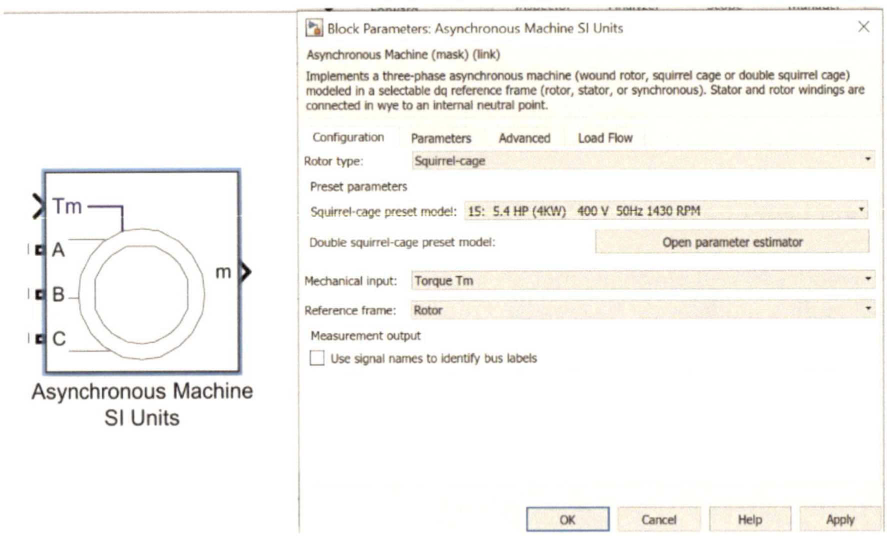

Figure 3.
Squirrel-cage asynchronous motor.

- Frequency: The motor should be operated at the nominal frequency specified by the manufacturer. Deviating from the nominal frequency can affect the motor's speed and torque characteristics and can also cause overheating and premature failure.

- Current: The motor should be operated within its rated current range. Overloading the motor can cause overheating and damage to the motor windings while operating the motor at low currents can result in reduced torque and efficiency.

- Cooling: The motor should be adequately cooled to prevent overheating. This can be achieved through natural convection, forced-air cooling, or liquid cooling, depending on the motor's design and operating conditions. In addition, it is important to ensure that the motor is properly installed, aligned, and coupled to the driven load. Regular maintenance and inspection can also help ensure that the motor continues to operate within its nominal parameters and provide reliable performance. **Figure 4** represents the starting technic of the Squirrel-cage asynchronous motor.

Ac voltage source 1: Peak amplitude: 326.59 phase = 0 frequency = 50

$$V_{ph} = \frac{V_{line}}{\sqrt{3}} = \frac{400}{\sqrt{3}} = 230.94 \quad (1)$$

$$V_{ph\_peak} = V_{ph\_RMS} * \sqrt{2} = 230.94 * \sqrt{2} = 326.59 \quad (2)$$

Ac voltage source 2: Peak amplitude: 326.59 phase = -120 frequency = 50
Ac voltage source 3: Peak amplitude: 326.59 phase = 120 frequency = 50
Tm: rated torque: use step; step time: 1 initial value: 0 final value $\frac{P[watt]}{\omega[ras/s]} = \frac{4000}{1430 * 2pi/60} = 26.7113[N.m]$

As shown in **Figure 5**, select the bus selector to choose output of the machine.

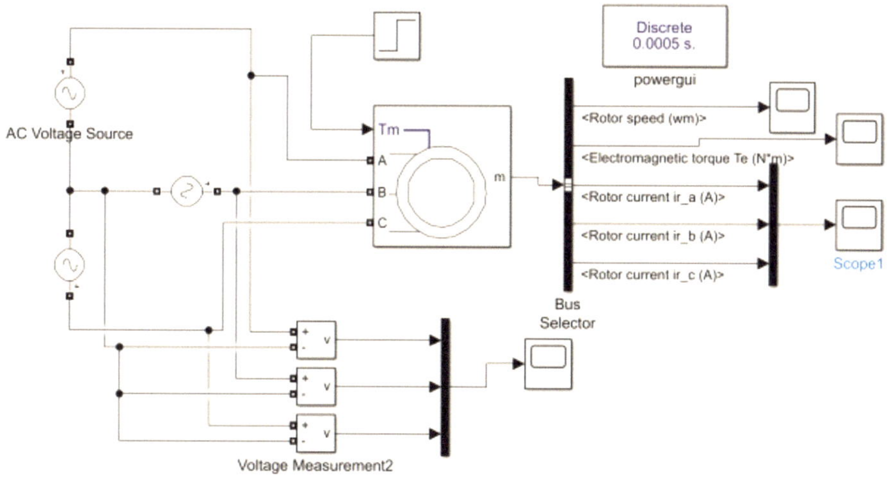

Figure 4.
Squirrel-cage asynchronous motor starting.

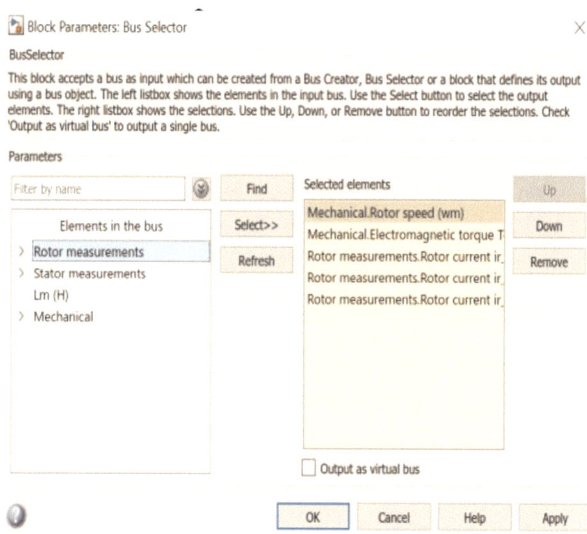

Figure 5.
Bus selector output.

3.2 Rated parameter three-phase asynchronous motor result

Run the simulation model, when the rated voltage is applied to the asynchronous motor, the motor begins to accelerate and increases its speed until it reaches its rated speed $\omega_{rated} \approx 150\ rad/s$. At this point, however, when a resisting torque is applied to the motor, such as when a load is attached to the motor's output shaft, the motor experiences a reduction in speed. When an asynchronous motor is first started, it draws a high current known as the starting current. The starting current is typically several times higher than the motor's rated current and is required to generate the initial torque to overcome the inertia of the motor and start the rotation of the motor's rotor.

As the load on the motor increases, the motor draws more current from the power supply to generate the required torque to overcome the resisting torque. This increased current causes a reduction in the motor's speed, which will continue to decrease as the load on the motor increases (**Figure 6**).

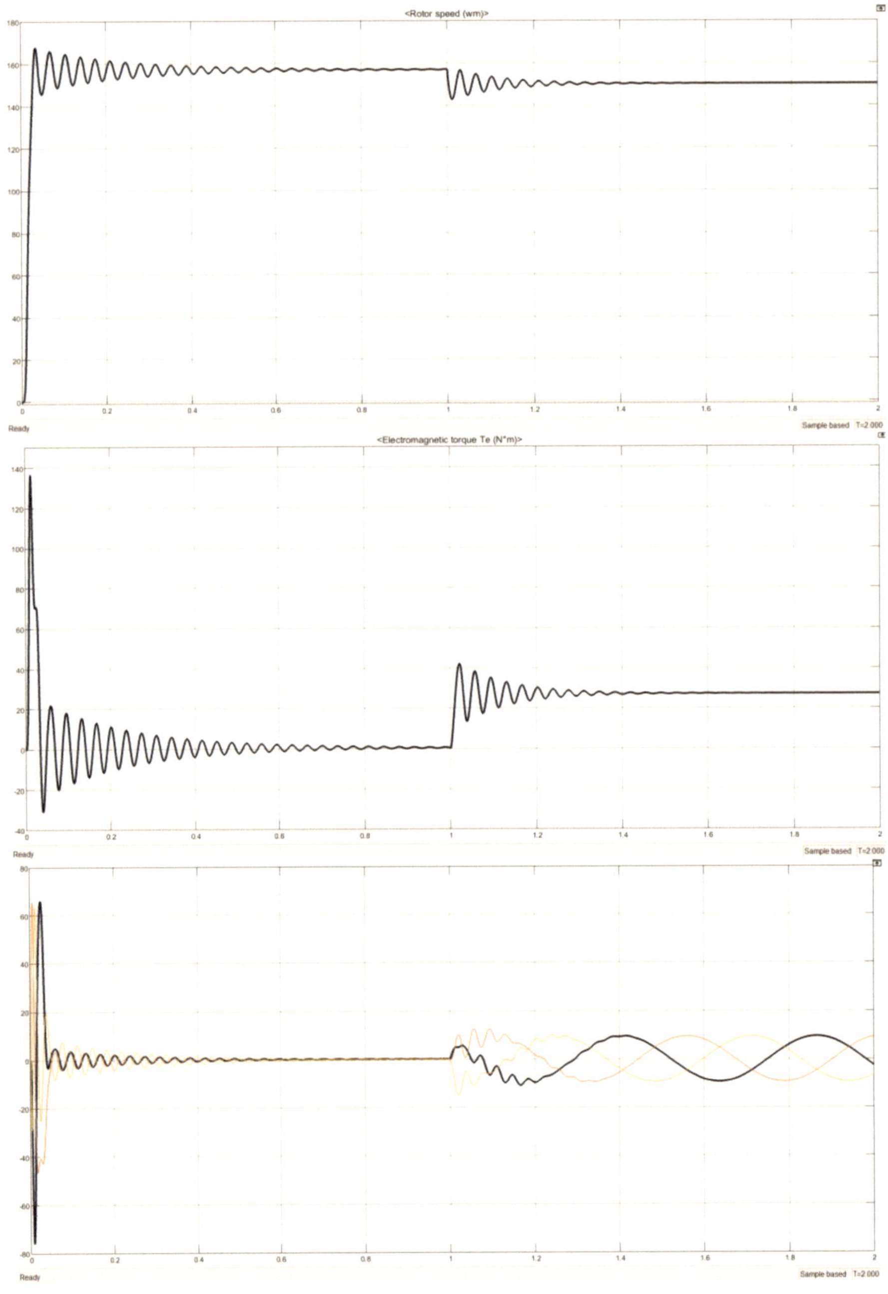

Figure 6.
Asynchronous motor output speed[rad/s], torque [N.m], and current [A].

4. Voltage control of the asynchronous motor

Voltage control of asynchronous motors using inverters is an effective technique for reducing starting current and improving the performance of the motor. However, it is important to consider the potential drawbacks and ensure that the system is properly designed and implemented to avoid any negative effects [8, 9].

Voltage control of asynchronous motors using inverters is a common technique used to reduce the starting current of the motor. In this method, the voltage supplied to the motor is controlled using an inverter to reduce the initial current surge that occurs during motor startup [10–14]. **Figures 7** and **8** represent the voltage control of an Asynchronous motor in Matlab Simpower system environment.

V_{dc} can be calculated as follows: $V_{ph\_peak} = m\frac{V_{DC}}{2}$, m is the efficiency so $V_{DC} = 650$ and for the PWM Generator (2-Level) choose three-phase bridges 6-pulses and carrier frequency $f_c = 100 * 50$.

The bloc repetitive sequence is responsible for the technique to reduce starting current; time value: [0 0.2 10], output value: [0 50 50], mi is the modulation index, and fr is the reference frequency.

Figure 9 shows that the motor speed responds quickly without a big overshot and for the starting current it does not exceed 25 A.

5. Linear V/f speed control of the asynchronous motor

Linear V/f speed control is another popular control technique for asynchronous motors. In this technique, the stator voltage and frequency are varied in a linear

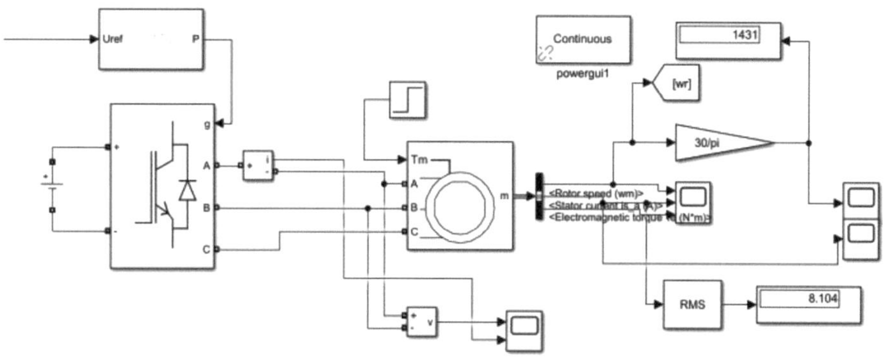

Figure 7.
Asynchronous motor, power part.

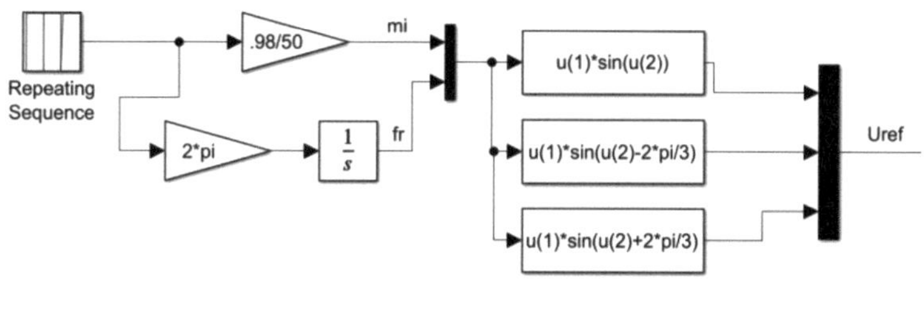

Figure 8.
Asynchronous motor voltage control, the control part.

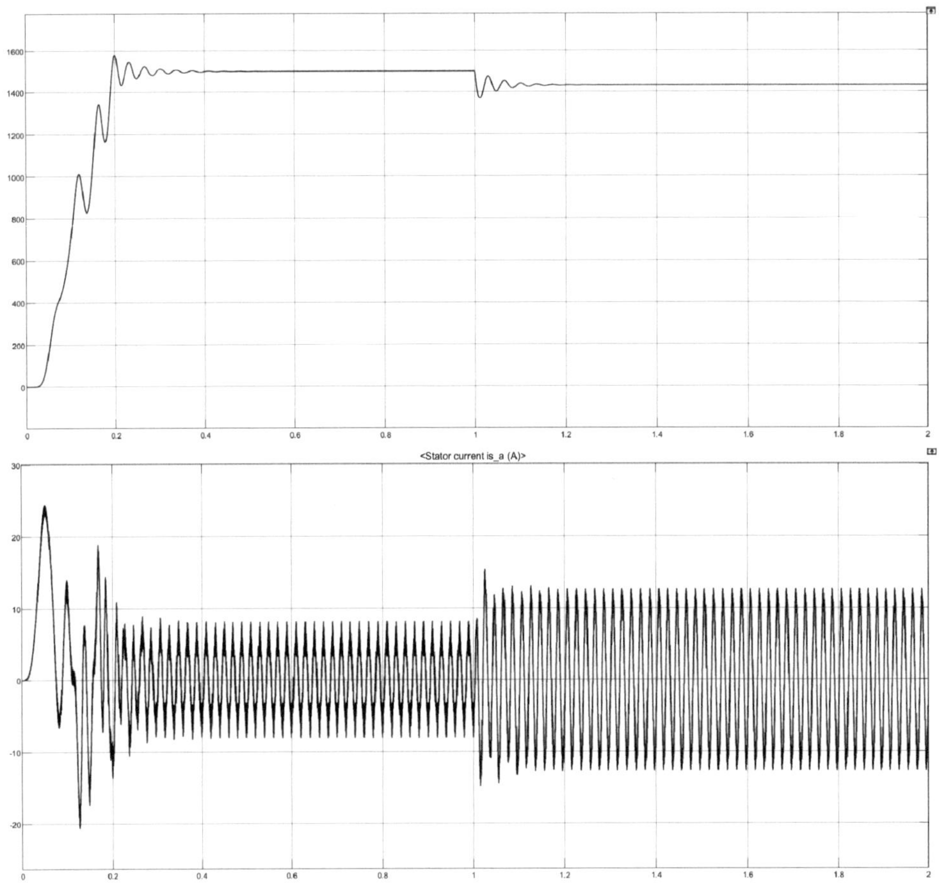

Figure 9.
Asynchronous motor output voltage control speed $[rad/s]$ *and current* $[A]$.

relationship with each other to maintain a constant ratio, known as the V/f ratio. The V/f ratio is proportional to the speed of the motor and is determined by the motor's parameters, such as the number of poles and the rated voltage and frequency [9, 15]. By controlling the V/f ratio, the speed of the motor can be controlled. Linear V/f control is a simple and cost-effective control strategy that is commonly used in applications where speed control precision is not critical, such as pumps, fans, and conveyors. However, it has limitations in terms of accuracy and efficiency, particularly at low speeds and under heavy loads. More advanced control techniques, such as field-oriented control, can provide better performance for a wider range of operating conditions. To implement linear V/f control, the stator voltage and frequency are controlled using the following relationship: $V/f = constant$ [9, 15–17].

To overcome the limitations of linear V/f control, voltage boosting can be used to extend the operating range of the motor. Voltage boosting refers to the practice of increasing the stator voltage above the rated voltage to improve the performance of the motor at high speeds and under heavy loads. The boosting voltage is notated by V_0

$$V = k * f + V_0 \text{ and } k = {}^{V_{rated}}\!/\!_{f_{rated}} \tag{3}$$

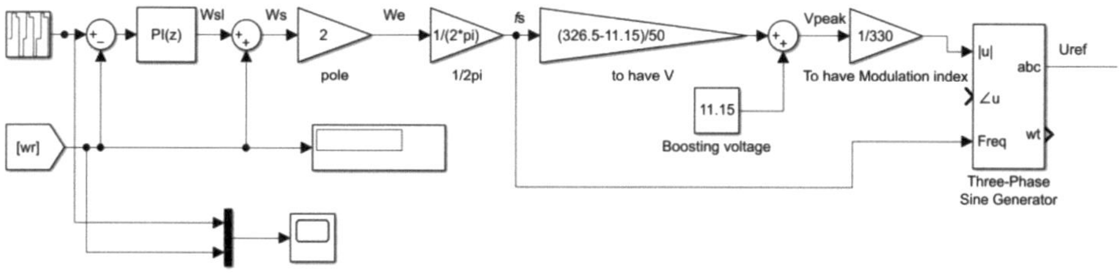

Figure 10.
Asynchronous motor linear V/f speed control, the control part.

From the torque equation, if the slip is small enough, the torque slip curve in the stable region can be approximated by [9, 15–17]:

$$T_e = \frac{3}{R_r}\left(\frac{V_s}{\omega_s}\right)^2(\omega_s - \omega_r) \tag{4}$$

$V_0 = I_s * R_s$ where I_s : Stator RMS current **Figure 10** and R_s Stator resistor

$$V_0 = 7.943 * 1.405 \tag{5}$$

Figure 10 shows the linear *V/f* speed control of the asynchronous motor. Repeating sequence represent reference speed in rad/s, the motor is driven in different speed references and in two directions: time value: [0 4 4 8 8 12 12 16 16] and output value: [100 100 120 120 –100 –100 –50 –50].

The output of the PI regulator represents the slip angular frequency ω_{sl} then ω_{sl} is added to the rotor speed ω_r to find the synchronous speed $\omega_r = \omega_r + \omega_{sl}$. To find the electrical angular frequency ω_e , ω_r is multiplied by the number of pole pairs $p = 2$.

To calculate $f_s = {}^{\omega_e}/_{2\pi}$; then V_{peak} is calculated by the equation $V = k * f + V_0$

PID controller P = 0.1 I = 1.82 (PID tuner method).

Figure 11 shows that the motor speed follows the reference speed perfectly, the motor responds quickly to the reference speed and reaches the speed of 100 rad/s in 2 s with a reduced overshoot of less than 30%, then when the reference speed changes

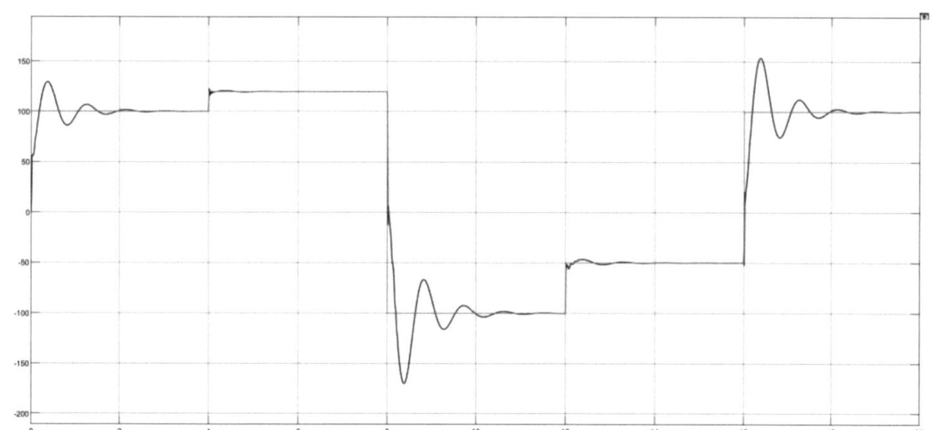

Figure 11.
Speed response asynchronous motor linear V/f speed control.

to 120 rad/s the time response is equal to 0.2 s for the overshoot that is less than 3%, but if the reference speed changes to the other direction to −100 rad/s the time response and the overshoot increase to 2.5 s and 70%, respectively, a same note like before when the speed changes on the same direction to −50 rad/s the time response is equal to 0.2 s for the overshoot of less than 3%, and the time response and the overshoot increase to 2.5 s and 50%, respectively, if the speed changes the direction to 100 rad/s.

6. Vector r V/f speed control of the asynchronous motor

Vector control, also known as field-oriented control (FOC), is a popular technique for controlling the speed and torque of asynchronous motors. Unlike V/F control, which controls the voltage and frequency applied to the motor, vector control directly controls the magnetic field and torque of the motor. This is achieved by decomposing the stator currents into two orthogonal components, known as the direct axis (Id) and quadrature axis (Iq) currents. The Id component represents the magnetizing current, while the Iq component represents the torque-producing current. By controlling the amplitude and phase of the Id and Iq currents, vector control can provide precise control of the motor's speed and torque, even at low speeds and under varying loads. Vector control can also improve efficiency and reduce the noise and vibration of the motor. However, vector control requires more complex hardware and software compared to V/F control, it is typically more expensive to implement [9, 18–23].

Vector control offers many benefits over other types of motor control, making it a popular choice for many applications requiring high-performance motor control; Advantages of vector control for asynchronous motors [9, 18–23]:

High performance: Vector control allows for high-performance control of the asynchronous motor, resulting in better torque control and faster dynamic response.

Wide range of speed control: Vector control can control the speed of the motor across a wide range, making it suitable for use in a variety of applications.

Improved Efficiency: Vector control allows for the motor to operate at optimal efficiency by minimizing losses in the rotor and stator.

Reduced harmonics: Vector control reduces the harmonic content in the motor current, which results in less vibration, noise, and electromagnetic interference.

Sensorless control: Vector control can be implemented using sensorless techniques, which eliminates the need for additional sensors and reduces the overall cost and complexity of the motor control system [9, 18–23].

Robustness: Vector control is robust to disturbances such as changes in load and supply voltage, making it suitable for use in harsh environments. **Figure 12** represents the block diagram of vector control.

As shown in the **Figures 13–15**, FOC subsystem represents the bloc that resume the vector control of Asynchronous motor it has three inputs and one output; the inputs are the speed reference ω_{r_ref}, the actual speed ω_r and the three phase current I_s. The speed reference is presented by a Repeating Sequence Stair, double click in the Repeating Sequence Stair and fill it by: Vector of output values: [100 120 –100 –80]. And Sample time: 4, The actual speed ω_r and the three phase current I_s are motor's output.

I_d reference is compared with I_d measured, the I_d reference is I_{s_peak} at the no-load test from **Figure 15**, $I_d = I_{s\_peak} = 6\,A$, and I_q reference is compared with I_q measured, I_{q_ref} is the output of the speed PI regulator,

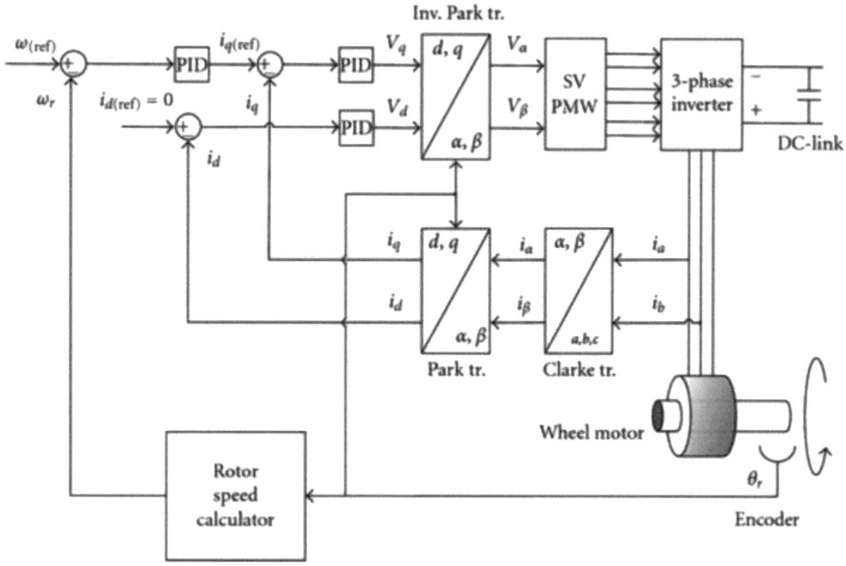

Figure 12.
Block diagram of vector control [23].

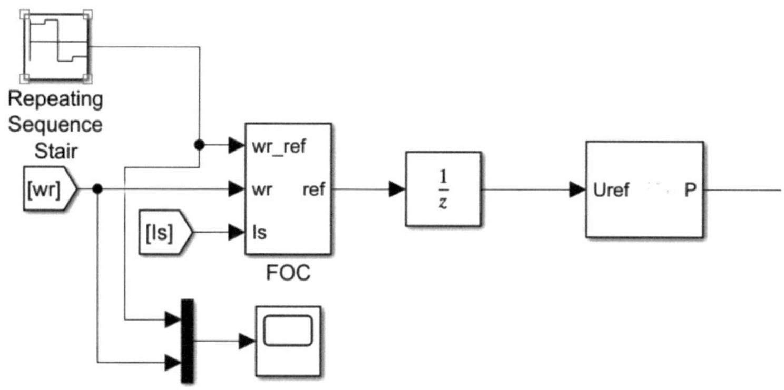

Figure 13.
Asynchronous motor FOC speed control, the control part.

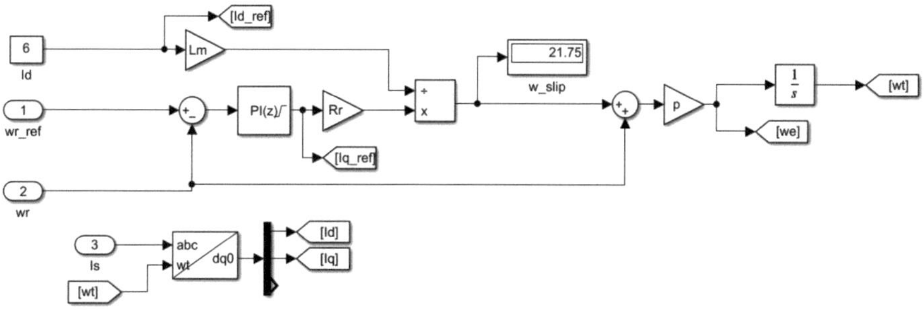

Figure 14.
Subsystem FOC speed control, part 1.

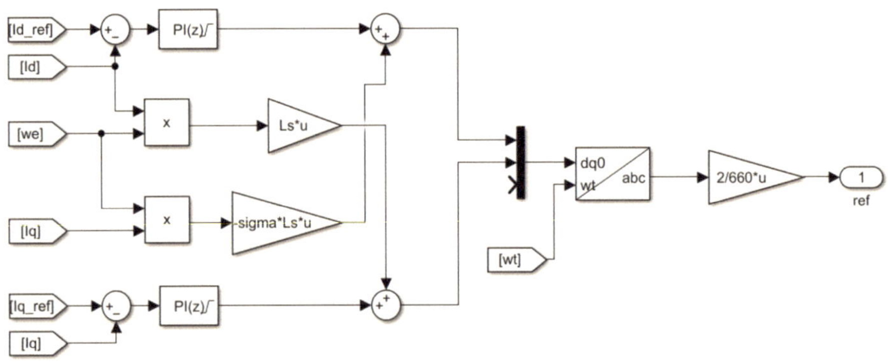

Figure 15.
Subsystem FOC speed control, part 2.

$\omega_{slip} = \frac{I_d L_m}{I_q R_r}$ and the rotor angle $\theta_{rotor} = \omega_e t = \omega_s p t = (\omega_{slip} + \omega_r) p t$

θ_{rotor} is used in park transformation to find I_d and I_q

$$V_{ds} = R_s I_{ds} + L_s \, p I_{ds} - sigma * \omega_s L_s I_{qs} \quad (6)$$

$$V_{qs} = R_s I_{qs} + sigma * L_s \, p I_{qs} + \omega_s L_s I_{ds} \quad (7)$$

The outputs of the currents PI regulators ensure the first term of the equation then we add $-sigma * \omega_s L_s I_{qs}$ to find V_{ds} and $\omega_s L_s I_{ds}$ to find V_{qs}, **Figure 16** shows the speed response asynchronous FOC speed control.

PI speed controller P = 0.6606563 I = 17.58084 (PID tuner method)

PI currents controller P = 14.39 I = 5800.047 (PID tuner method)

Figure 16 shows that the motor speed follows the reference speed perfectly; the motor responds quickly to the reference speed and reaches the speed of 100 rad/s in 0.4 s with a reduced overshoot of less than 9%, then when the reference speed changes to 120 rad/s the time response is equal to 0.4 s for the overshoot that is less than 0.75%, but if the reference speed changes to the other direction to −100 rad/s the time response and the overshoot increase to 0.7 s and 13%, respectively, after applying of a

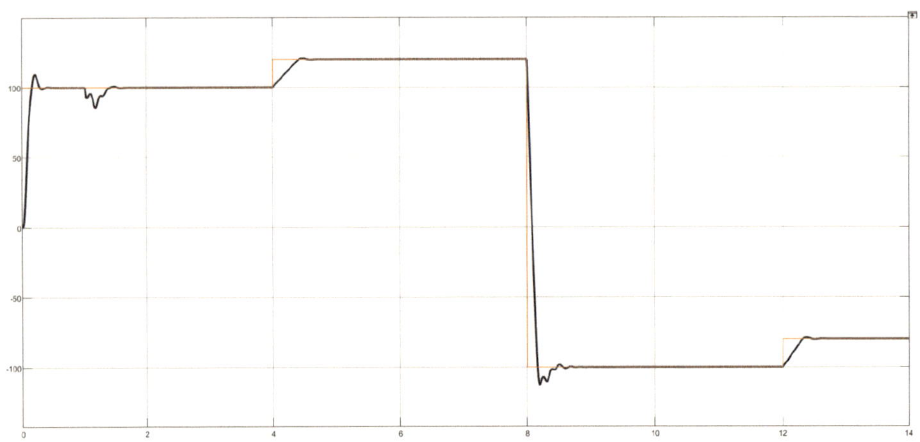

Figure 16.
Speed response asynchronous FOC speed control.

rated torque at 1 s a small disturbance is introduced on the response of the speed then the disturbance is rejected to return to the pursuit of the reference speed.

7. Conclusions

MATLAB SimPowerSystems was used to model and evaluate the performance of asynchronous motors under various control schemes. According to the simulation results, vector control is more efficient and offers superior dynamic performance than voltage control and V/F control. It is also important to note that the usage of SimPowerSystems in this study gives the simulation additional realism, which can then be immediately used in the construction of control systems utilizing hardware-in-the-loop simulation platforms such as dSPACE. This work offers important insights into how asynchronous motor performance changes with various control schemes, and it can be used as a useful resource for further study in this area.

Each of the three asynchronous motor control methods—voltage control, V/F control, and vector control—has specific benefits and drawbacks. Voltage control is straightforward and simple to use, but it has limitations when it comes to high performance and dynamic reactions. Although V/F control offers superior torque control and better dynamic response, it has problems with speed accuracy and stability. While improving torque response and speed stability, vector control provides the highest level of performance and control accuracy.

It should be emphasized that compared to the other two control systems, vector control is more expensive and challenging to execute because it requires more sophisticated hardware and software. Vector control has recently become more affordable and accessible because of technological improvements.

As a result, the selection of a control strategy for the operation of an asynchronous motor will be based on the particular needs of the application, taking into account elements such as complexity, cost, and performance. Overall, it is evident that vector control provides the best performance and control accuracy, making it the go-to choice for high-performance applications that demand exact torque and speed control. However, in less demanding applications where simplicity and cost-effectiveness are more crucial, voltage control and V/F control still have their place.

Author details

Fezazi Omar[1]* and Ayad Ahmed Nour el Islam[2]

1 ICEPS Laboratory (Intelligent Control & Electrical Power Systems), Djillali Liabes University, Sidi Bel-Abbes, Algeria

2 Kasdi Merbah University Ouargla, Algeria

*Address all correspondence to: fezaziomarfz@gmail.com

References

[1] You Electrical Guide. Three Phase Induction Motor - Working Principle. Your Electrical Guide. 2017. [Online]. Available: https://www.yourelectricalguide.com/2017/07/three-phase-induction-motor-working-principle.html

[2] Fitzgerald AE, Kingsley Jr. C, Umans SD. Electric Machinery. New York: McGraw-Hill Encyclopedia of Science and Technology; 2003

[3] Wildi T. Electrical Machines, Drives, and Power Systems. New Jersey, U.S.: Prentice Hall; 2006

[4] Boldea I, Nasar SA. The Induction Machine Handbook. Boca Raton, Florida: CRC Press; 2010

[5] Miller TJE. Induction Motors. Boca Raton, Florida: CRC Press; 1993

[6] Patil HB, Mulay YS. Design and development of three-phase induction motor. International Journal of Engineering Research & Technology. 2015;**4**(6):1130-1135

[7] Zhang Y, Wang D. Modeling and simulation of AC motor drive systems using SimPowerSystems. IEEE Transactions on Industrial Electronics. 2017;**64**(7):5693-5703. DOI: 10.1109/TIE.2017.2661527

[8] Zhang Y, Wang D. Hardware-in-the-loop simulation of power electronics and electric machines using SimPower Systems. IEEE Transactions on Industrial Electronics. 2019;**66**(3):1983-1993. DOI: 10.1109/TIE.2018.2823105

[9] Novotny DW, Lipo TA. Induction motor vector control: A tutorial. In: Vector Control and Dynamics of AC Drives. Oxford University Press; 2008. pp. 171-219

[10] Ogata K. Modern control engineering. New Jersey, U.S.: Prentice-Hall; 1997

[11] Depenbrock M. Direct self-control (DSC) of inverter-fed induction machine. IEEE Transactions on Power Electronics. 1988;**3**(4):420-429. DOI: 10.1109/63.192770

[12] Rodriguez J, Pontt J, Correa P, Rebolledo R. Review of control strategies for improving dynamic performance of induction motors. IEEE Transactions on Industrial Electronics. 2007;**54**(6): 2934-2945. DOI: 10.1109/TIE.2007.908302

[13] Ardebili M, Iravani R. Dynamic voltage restorer (DVR) compensation of voltage disturbances using synchronous-reference-frame control. IEEE Transactions on Power Delivery. 2004; **19**(1):405-413. DOI: 10.1109/TPWRD.2003.820838

[14] Bolognani S, Tani A. An innovative low-cost sensorless vector control for induction motor drives. IEEE Transactions on Industrial Electronics. 2010;**57**(8):2752-2762. DOI: 10.1109/TIE.2009.2037208

[15] Bose BK. Modern power electronics and AC drives. New Jersey, U.S.: Prentice Hall; 2002

[16] Krishnan R. Modeling and control of induction motor drives. Hoboken, NJ, United States: John Wiley & Sons; 2001

[17] Krishnan R. Electric Motor Drives: Modeling, Analysis, and Control. Vol. 32. Pearson Education India; 2010

[18] Leonhard W. Vector control of three-phase AC machines: System development in the practice. IEEE

Industrial Electronics Magazine. 1996; **10**(2):26-33. DOI: 10.1109/57.491447

[19] Holtz J. Field-oriented control of AC machines. IEEE Transactions on Industrial Electronics. 2001;**48**(4):708-719. DOI: 10.1109/41.956193

[20] Novotny DW, Lipo TA. Vector control and dynamics of AC drives. England: Oxford University Press; 2008

[21] Botto MA. Advanced control of AC/DC power networks: System of systems approach based on spatio-temporal scales. In: In 2016 18th European Conference on Power Electronics and Applications (EPE'16 ECCE Europe). IEEE; 2016. pp. 1-10

[22] Noroozian R, Jafarzadeh M. Vector Control of Asynchronous Motor. London, UK, London, UK: INTECH; 2010. DOI: 10.5772/8673 Available from: https://www.intechopen.com/books/electric_motors_and_dr€ives/vector-control-of-asynchronous-motor

[23] Oliveira JG, Schettino H, Gama V, Carvalho R, Bernhoff H. Implementation and control of an AC/DC/AC converter for double wound flywheel application. Advances in Power Electronics. 2012; **2012**:604703. DOI: 10.1155/2012/604703

Chapter 5

Dynamic Analysis and Optimized Design of Synergetic Control for a PMSM Drive System

Andrew Adib and Rached Dhaouadi

Abstract

This chapter presents an optimum design of synergetic control for a permanent magnet synchronous motor (PMSM) drive system. New macro-variables are proposed to improve the performance of the standard controller. The controller's performance is compared with that of the field-oriented control scheme. The chapter also investigates the regenerative braking mode of operation in PMSM. Regenerative braking is achieved by operating the motor in torque control mode. The different algorithms are validated through experiments using a 1-hp PMSM drive system. We also provide an extensive study of the controller parameters tuning for optimal performance. The experimental results show that the proposed macro-variables improve the performance of the synergetic controller significantly. The synergetic controller is able to overcome nonlinearities in the system, such as static friction, faster than the field-oriented controller. The system also experiences fewer harmonics under the synergetic controller. The synergetic controller shows also better performance under wide signal variations. As for regenerative braking, the torque control mode of operation is shown to be suitable for harvesting energy and both techniques showed similar performance levels. The proposed synergetic control strategy will be very useful in electric vehicle (EV) applications, as it allows to improve the dynamic response and efficiency of the drive system required by the EV dynamics.

Keywords: synergetic control, field-oriented control, permanent magnet synchronous motor, regenerative braking, torque control

1. Introduction

Electric vehicles are gaining increasing attention by automotive manufacturers in an attempt to reduce carbon footprints and produce more eco-friendly vehicles. Electric vehicles are powered by electric motor drives instead of the conventional combustion engine. Permanent Magnet Synchronous Motor (PMSM) drives have received a lot of interest in this field for their powerful torque profiles that are suitable for vehicle traction [1–6].

Regenerative braking is crucial for electric vehicle applications since the mechanical energy of the motor during deceleration or braking periods can be harvested and used to charge the vehicle's battery or contained within an energy storage unit designed for such purpose. In electrical braking, the kinetic energy of the motor can be harvested, with the proper switching sequence of the three-phase inverter, in order to redirect the current flowing in the motor windings back to the source or to an integrated hybrid energy storage system. It should be noted that not all energy can be recovered during the braking process as there exist some boundaries along the torque-speed plane where regenerative braking occurs [7]. In most applications, this regenerated current charges a supercapacitor module, or the vehicle's battery, connected to the DC side of the inverter through a DC-DC converter. For AC motor drives, the power switches of the three-phase inverter can be utilized to harvest mechanical energy and feed it back to the DC source [8, 9]. The three-phase inverter works as a bidirectional DC-AC converter that controls energy flow in both directions between the power source and the motor in the case of both driving and regenerative braking modes of operation.

Several control techniques are available for controlling AC motor drives, such as vector control and direct torque control (DTC). Vector control, also known as field-oriented control (FOC), is the most commonly utilized control technique in AC motor drives. It is a reliable control scheme in most drive applications. The main role of field orientation is to decouple the machine's generated flux from the torque in order to control both parameters separately as desired. This is done by replacing the three-phase stationary representation of the stator field with a dynamic two-phase rotating d-q frame, where the rotor magnetic field is aligned with the d-axis. In this process, the motor torque can be controlled separately by controlling the current in the quadrature (q) axis. Similarly, the flux is controlled independently by controlling the current in the direct (d) axis. Operating a three-phase AC machine under field orientation is similar to the operation of a separately excited DC machine where the field circuit controls the generated flux, while the armature circuit regulates the torque of the motor [3].

The FOC scheme consists of two control loops: an inner loop consisting of current regulators and an outer loop controlling the speed of the motor. In general, the torque or current regulators have to be tuned before tuning the speed controller, and the speed controller cannot be tuned without the assumption that the current regulators are tuned perfectly [1, 2].

Unlike FOC, DTC does not require any current regulator, coordinate transformation and PWM signals generator [3, 10]. In spite of its simplicity, DTC allows a good torque control in steady state and transient operating conditions. In addition, this controller is very little sensible to the parameters detuning in comparison with FOC. The main principle of DTC is to adjust the stator voltage vector based on the errors between the actual torque and stator flux linkage as well as their references. This method requires stator resistances be known and precludes both current controllers and motor parameters. Hence, it is insensitive to parameter variations of the machine. DTC also provides fast torque responses and simple implementation. However, DTC is known for its high torque and flux ripples that lead to vibrations and increase of losses [11–13].

On the other hand, it is acknowledged that DTC and FOC presents some disadvantages that can be summarized in the following points: 1) difficulty to control torque and flux at very low speed; 2) high current and torque ripple; 3) variable switching frequency behavior; 4) high noise level at low speed; 5) sensitivity to the stator resistance, rotor time constant, or mutual inductance parameters.

Various control strategies have been also proposed in the literature to solve these problems for PMSM drives and improve the robustness to modeling or parametric uncertainties, such as sliding mode control, adaptive control, predictive control, neural networks, as well as hybrid control methods [14–17]. It is recognized that these different strategies have their own merits, and all meet the requirements of modern high-performance drives. However, most techniques are found to be computationally intensive, and show some performance limitations in terms of flexibility, implementation, and required bandwidth.

Therefore, new control techniques have to be developed to improve the performance of PMSM drives. It is essential to develop advanced control strategies for power conversion systems so that the robustness, flexibility, and dynamic performance of PMSM can be significantly improved.

Synergetic control (SC) is a nonlinear control technique that is well-suited for nonlinear systems. It overcomes the disadvantages of the previous controllers for PMSM, and linear model-based controllers who do not perform well with nonlinear plants under large signal variations [18–20]. The theory of synergetic control, however, overcomes these disadvantages via the use of all of its nonlinearity in the controller design process. This control theory is close to that of sliding mode controllers; both theories share similar properties and advantages, such as order reduction of the system and decoupling design procedure. Despite this, SC is still more advantageous compared to the sliding mode control primarily because it does not require high bandwidth to operate, as is the case with the sliding mode control. This makes it suitable for digital implementation. Furthermore, contrary to sliding mode controllers, SC maintains constant switching frequencies which lead to less noise due to switching in the system. Finally, synergetic control operates continuously as opposed to the sliding mode controller, which is known for its discontinuity. As previously mentioned, the system model is used for the controller synthesis process. Although this leads to better control as it provides more information about the system dynamics, it also makes the controller sensitive to errors in system parameters considered in the model itself. However, this can be overcome by proper selection of the macro-variables which is further discussed in Section 3.

Synergetic control has been previously used in motor drive applications [14, 15, 21–24]. It has been implemented as a nonlinear speed control of multiple induction motor (IM) and PMSM drives. In [21], the authors derive a control law that reduces the order of the system and devise a simple method to select the closed loop poles of the system. However, they do not take into account any load variations applied on the system, which will affect the steady state speed of the drive. On the other hand, researchers in [22] introduce an adaptive controller based on the synergetic approach to control a PMSM drive. Although this controller takes into consideration any application of load torque on the motor, there remains some steady state error in the motor speed. In [23], the authors implement a synergetic controller to control the speed of an IM drive. They show the effect of changing some controller parameters on the response speed, transient behavior, steady state error and response to load applications. However, they propose excessive numbers of manifolds for a two control input system. In [24], the authors use both synergetic control and vector control to control an IM drive. The proposed controller uses the synergetic control theory to control the speed and generate reference d-q currents. The motor currents are regulated using vector control and hysteresis current control. In [14], the authors compare between the synergetic and the sliding mode controllers in controlling the speed of an IM. The results show that the synergetic controller was able to reduce the effect of load torque application as well as vibrational torques caused by the discontinuities presented by the sliding mode controller. In [15], three adaptation based nonlinear controllers namely adaptive terminal sliding mode, adaptive terminal synergetic and adaptive synergetic controllers have been used for the unified model of fuel cell, battery and ultracapacitor based hybrid electric vehicle to regulate the output voltage of the DC bus and to cater for the parametric variations of the system.

In this paper an optimized synergetic controller has been applied for the speed control of a PMSM drive. The particularity of the method is the incremental design and optimization of the macro-variables to improve the performance of the SC. The new macro-variables enhance the behavior of the d-axis current and eliminate any steady-state errors. The performance of this controller is improved compared to the standard controller used in the literature. A special macro-variable is also designed and implemented to control the torque and operate the motor under regenerative braking.

The synergetic controller is designed and implemented on a 1-hp prototype setup. FOC is also implemented and used as a reference to test the performance of the proposed synergetic controller under different operating conditions, such as response time, response to load torque variations, and regenerative braking. Regenerative braking under FOC and SC is also investigated and implemented. The synergetic controller showed very good robustness against disturbances with a reduced level of harmonics in the system. The proposed control scheme has also the characteristics of finite time convergence and chattering free phenomena.

2. PMSM dq model

In this paper, we consider a three-phase surface mounted PMSM with sinusoidal stator windings distribution. In order to simplify the analysis of the PMSM, the mathematical model is transformed into a two-phase reference frame using the Clarke-Park transformation [1]. This transforms the voltages and currents from a three-phase reference frame (uvw) to a fixed two-phase reference frame (αβ) as illustrated in **Figure 1**. The PMSM stator consists of three-phase windings (abc).

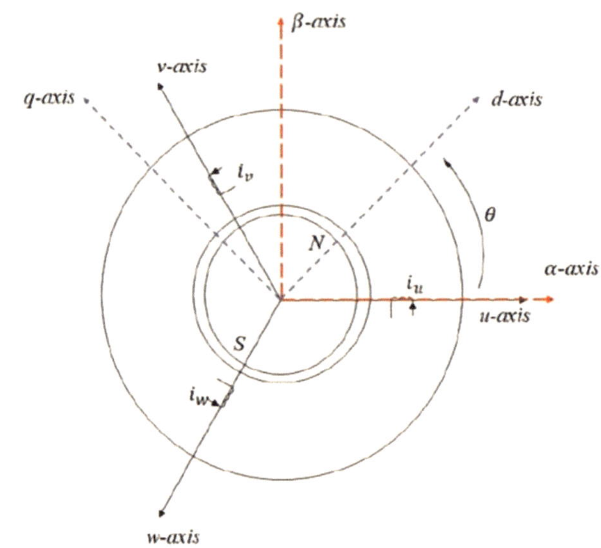

Figure 1.
PMSM reference frames.

The (αβ) reference frame is fixed to the stator and the (dq) frame is rotating with an angular speed ω equal to the synchronous speed. The permanent magnets are surface mounted on the rotor and thus create a rotating magnetic field. Under field orientation, the PMSM model in the rotor reference frame is given by

$$v_d = Ri_d + L_d \frac{di_d}{dt} - L_q i_q \omega_e \tag{1}$$

$$v_q = Ri_q + L_q \frac{di_q}{dt} + L_d i_d \omega_e + \omega_e \lambda_{PM}, \tag{2}$$

$$J\frac{d\omega}{dt} = T_{em} - B\omega - T_L \tag{3}$$

$$T_e = \frac{3}{2}\frac{P}{2}\left[\lambda_{PM} i_q + \left(L_d - L_q\right) i_d i_q\right] \tag{4}$$

where $\left(v_d, v_q, i_d, i_q\right)$ are the dq components of voltages and currents in the rotating dq frame, R is the stator resistance per phase, $\left(L_d, L_q\right)$ are the direct and quadrature self-inductances, λ_{PM} is the flux produced by the permanent magnets on the rotor, T_{em} is the generated electromagnetic torque, and P is the number of poles. For the case of surface mounted or non-salient PMSM, the inductances in both d and q axes are equal.

3. Synergetic control strategy

3.1 Synergetic control law formulation

The synergetic control design procedure is based on the Analytical Design of Aggregated Regulators (ADAR) method [18, 19, 25]. We consider a non-linear system, described by its state space equation in the form:

$$\dot{x} = f(x, u, t) \tag{5}$$

where, x is the state vector defining all the state variables of the system, u is the control input vector, and t denotes time. First, a macro-variable is defined as a function of the system state variables

$$\psi = \psi(x) \tag{6}$$

The objective of the control effort is to force the system dynamics to operate on the manifold $\psi = 0$. Defining a manifold adds a new constraint on the system and hence reduces the order of the system. This will also force the system towards stability, preventing it from diverging. The macro-variable should be carefully selected to achieve the control specifications and limitations, such as avoiding saturation. A macro-variable is defined for each desired control input and its dynamics are subjected to the following constraint

$$T\dot{\psi} + \psi = 0, \quad T > 0 \tag{7}$$

where T is a time constant that defines the convergence speed to the manifold specified by the macro-variable.

T can be tuned and selected by the designer. Defining the appropriate macro-variable will force the system towards global stability. As discussed earlier, the controller synthesis depends greatly on the system model. Therefore, any uncertainties in the model parameters may affect the controller's performance. To avoid the use of costly complex observers in estimating system parameters, studies show this drawback can be overcome by appropriate selection of the macro-variables [25].

By using the chain rule of differentiation, the expression of and the constraint equation are computed as

$$\frac{d\psi}{dt} = \frac{d\psi}{dx} \times \frac{dx}{dt} \tag{8}$$

$$T\frac{d\psi}{dt}f(x,u,t) + \psi = 0 \tag{9}$$

3.2 Conventional synergetic controller design for the PMSM

The PMSM model is first written in state space form by selecting the dq current components (id, iq) and the speed ω as state variables. Eqs. (1)–(2) are next as follows:

$$\frac{di_d}{dt} = \frac{1}{L_d}\left[v_d - Ri_d + L_q i_q \omega_e\right] \tag{10}$$

$$\frac{di_q}{dt} = \frac{1}{L_q}\left[v_q - Ri_q - L_d i_d \omega_e - \omega_e \lambda_{PM}\right] \tag{11}$$

It can be seen from this representation that the PMSM has two control input channels: vd and vq. Thus, two macro-variables are needed. A macro-variable is defined for each axis, which also means that torque and flux will be decoupled similar to FOC. In the literature, the following set of macro-variables are used [21].

$$\psi_1 = i_d \tag{12}$$

$$\psi_2 = K_1(\omega - \omega_{ref}) + K_2 i_q + K_3\int(\omega - \omega_{ref})dt \tag{13}$$

where K1, K2, K3 are the SC gains. The control effort forces the macro-variable to operate on the manifold $\psi = 0$, following the dynamics given by Eq. (7). Therefore, the first macro-variable ψ_1 is selected to be equal to the direct-axis current i_d since it is preferable to keep $i_d = 0$ in PMSM systems to avoid any flux weakening or demagnetization of the permanent magnets. As for the second macro-variable ψ_2, the first and last terms act as a speed controller and, together, they generate a command reference for the current in the q-axis i_q. ω_{ref} is the reference speed set by the designer. By selecting $K_2 = -1$, the resultant will be the error in the q-axis current $i_q^* - i_q$, which converges to zero automatically following the dynamics of the macro-variable.

To derive the control effort, the constraint given by Eq. (7) is applied for each macro-variable. The system Eq. (10)–(13) are next combined and solved to find the control input voltages as follows:

$$v_d^* = \left(R - \frac{L_d}{T_d}\right)i_d - L_q\omega_e i_q, \tag{14}$$

$$v_q^* = -\frac{K_1 L_q}{K_2 T_q}(\omega - \omega_{ref}) - \frac{L_q}{T_q} i_q - \frac{K_3 L_q}{K_2 T_q}\int(\omega - \omega_{ref})dt - \frac{K_1}{K_2} L_q \frac{d\omega}{dt} - \frac{K_3}{K_2} L_q(\omega - \omega_{ref}) + R i_q + L_d \omega_e i_d + \omega_e \lambda_{PM}. \quad (15)$$

As seen from Eqs. (14) and (15), the expression for vd and vq are a function of the system parameters and the variables Td and Tq, which are selected by the designer. The controller gains K1, K2, K3 are also tuned and selected by the designer. In conclusion, the synergetic controller requires four inputs; desired motor speed ω_{ref}, dq currents (id, iq), and actual motor speed. The expressions for each of the two control inputs are computed and the commands voltages (vd, vq) are computed. These voltages are then converted into three phase voltages using the inverse Clarke-Parke transformation and the corresponding SPWM signals are generated to control the switching of the VSI as desired. **Figure 2** shows a block diagram representation of the synergetic controller.

3.3 Proposed synergetic controller

The selected macro-variable ψ_1, given by Eq. (12), was found to lead to steady-state errors in the d-axis variables. Therefore, to eliminate any steady-state errors, the following modified macro-variable is proposed to improve the performance and maintain a constant value for the d-axis current $i_d = 0$.

$$\psi_{1m} = K_1 i_d + K_2 \int i_d dt. \quad (16)$$

$$\psi_2 = K_3(\omega - \omega_{ref}) + K_4 i_q + K_5 \int(\omega - \omega_{ref})dt \quad (17)$$

The introduction of the gain K1 in the macro-variable ψ_{1m} will control the peak magnitude of id during the transient periods. The addition of an integral term removes any steady state error and the gain K2 controls the speed of convergence during the transient periods. The macro-variable ψ_2 controlling the q-axis command voltage vq remains unchanged similar to Eq. (13). Only the gain subscripts have been updated.

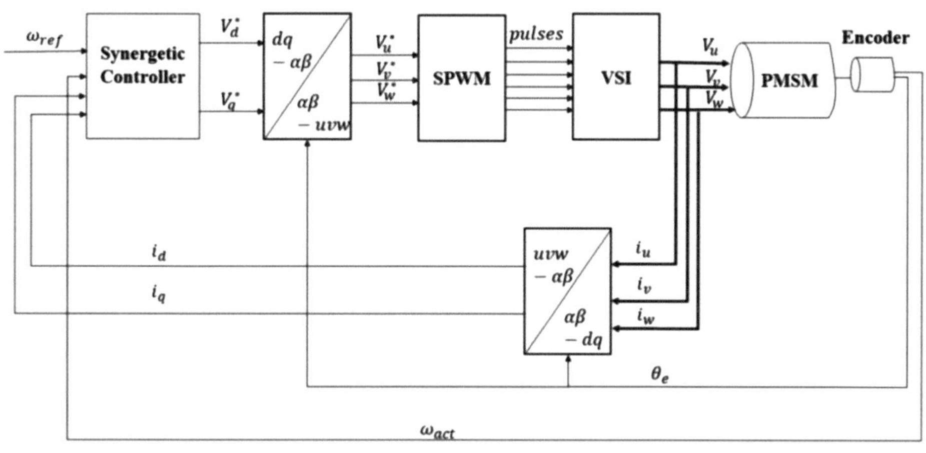

Figure 2.
Synergetic controller block diagram.

Next, the same steps from Section 3-B are followed to compute the new control law based on the proposed macro-variable in Eq. (16). The new control voltage is computed as follows:

$$v_d^* = -\frac{L_d}{T_d} i_d - \frac{K_2}{K_1}\frac{L_d}{T_d}\int i_d dt - \frac{K_2}{K_1} L_d i_d + R i_d - L_q i_q \omega_e \tag{18}$$

This command effectively maintains the d-axis current i_d at zero despite any discrepancies between the actual and system model parameters.

4. Regenerative braking strategy

4.1 Regenerative braking using FOC

For the PMSM model, regenerative braking (RB) is achieved by operating the motor in torque control mode, under FOC. RB analysis is next performed to determine the lower and upper boundaries of the applied torques that identify the RB region on the speed-torque plane. Through balancing the input and output power, the following expression for the input power can be formulated to account for power losses in the windings.

$$P_{in} = \frac{3P}{4} \lambda_{PM} i_q \omega + R\left(i_d^2 + i_q^2\right) \tag{19}$$

To find the regenerative braking boundaries, Pin is set to 0 indicating that no power is drawn from or supplied to the DC source. Next, the equivalent electromagnetic torque is found and minimized with respect to id.

$$T_{em} = 0 \ \ or \ \ T_{em} = -\frac{\omega}{R}\left(\frac{3P}{4}\lambda_{PM}\right)^2 \tag{20}$$

Then, the local minimum points to minimize the input power to the motor are found, and the maximum current absorbed by the DC source during regenerative braking can be calculated as follows.

$$\begin{aligned} i_d &= 0, \\ i_q &= -\frac{3P}{8R}\lambda_{PM}\omega. \end{aligned} \tag{21}$$

To achieve RB under FOC, the current controllers are given the command signals (i_d, i_q) equal to the results found in Eq. (20). This will force the motor to operate in the RB region and guarantee maximum current harvested from the motor's rotation. For that reason, the speed control loop is removed from the FOC controller and instead, the current commands (i_d, i_q) are fed directly to the current regulators. **Figure 3** shows the block diagram of the FOC scheme in RB mode.

4.2 Proposed regenerative braking using SC

Two macro-variables are defined for the synergetic controller to operate the PMSM in regenerative braking mode. The design of these macro-variables adopts the

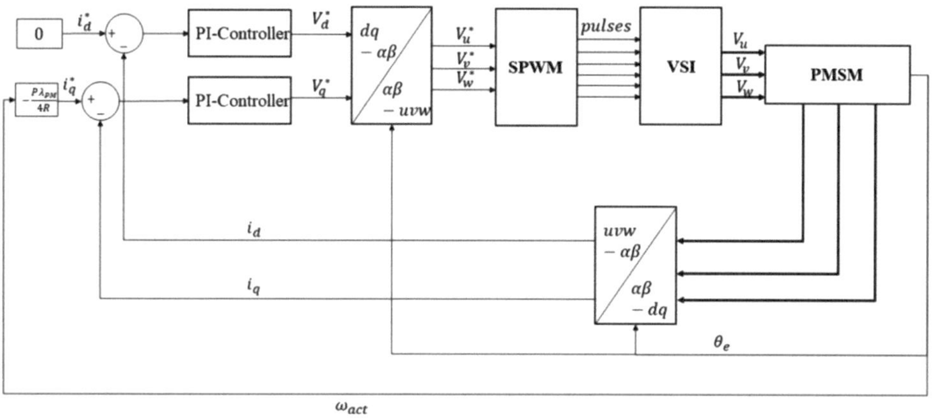

Figure 3.
FOC scheme in regenerative braking mode.

same approach to control the currents (i_d, i_q) and force them to follow their references described in Eq. (20). **Figure 4** shows the block diagram of the synergetic controller in regenerative braking mode. The proposed SC d-axis macro-variable ψ1sc is described by the same Eq. (16) used previously for the speed control scheme (ψ1sc = ψ1m). The reference d-axis current i_d^* is set to 0. The gain K4 is used to force id to zero and reduce the error between id and its reference. The integral term is added to eliminate any steady-state error.

The dynamic response of this macro-variable is governed by Eq. (7), where Td is the convergence time of ψ1. Following the design procedure explained in section III, the control law vd is derived as illustrated in Eq. (17). Applying this voltage vd will ensure that id remains at 0.

The proposed SC q-axis macro-variable ψ2sc to control iq is defined as

$$\psi_{2sc} = K_6\left(i_q - i_q{}^*\right) + K_7 \int \left(i_q - i_q{}^*\right) dt \qquad (22)$$

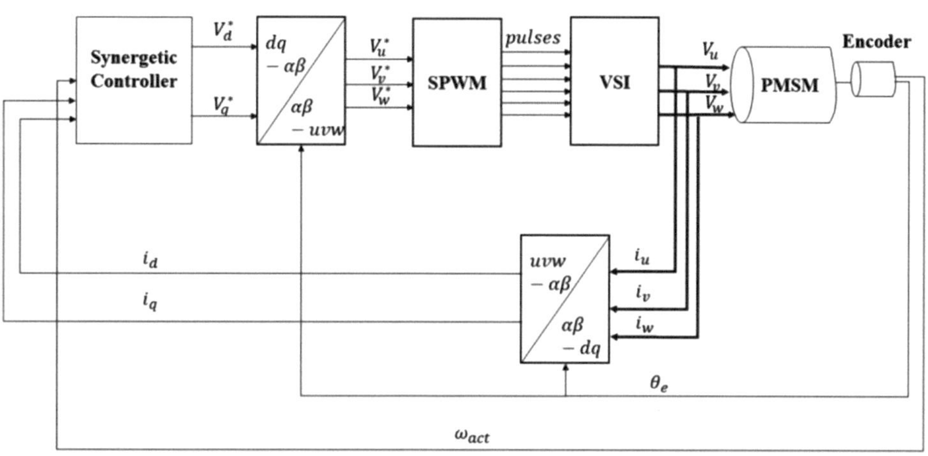

Figure 4.
Synergetic controller in regenerative braking mode.

where, i_q^* is the reference q-axis current that allows maximum current absorption by the DC supply as given by Eq. (20). The dynamic response of this new macro-variable will also be governed by Eq. (7), where T_q is the convergence time.

To evaluate the voltage command v_q^*, the design procedure explained in Section 3 is followed to get:

$$v_q^* = L_q \frac{di_q^*}{dt} - \frac{K_7}{K_6} L_q \left(i_q - i_q^*\right) - \frac{L_q}{T_q}\left(i_q - i_q^*\right) - \frac{K_7}{K_6}\frac{L_q}{T_q}\int\left(i_q - i_q^*\right)dt + Ri_q + L_d i_d \omega_e + \omega_e \lambda_{PM}. \tag{23}$$

Applying this voltage will generate a reverse torque that forces the motor to operate under regenerative braking mode.

5. Experimental analysis of synergetic control

5.1 Test bench description

Figure 5 shows a picture of the PMSM test bench. The system consists of a PMSM supplied with a voltage source inverter Myway MWINV-9R144 [26]. The inverter switches are controlled using the dSPACE 1103 board. The PMSM is coupled with a Bühler DC Motor through flexible couplings and additional disc inertia mounted on the same shaft. The DC motor acts as a mechanical load and is controlled using a DC-DC converter, which in turn is controlled using a dSPACE 1104 board. Two encoders are used: one incremental encoder is directly coupled to the DC motor side, while the other sine/cosine encoder is connected to the PMSM side. The analog encoder provides high resolution and accurate measurements necessary for field orientation in FOC. An interface board is designed to interface the analog encoder with dSPACE 1103. The PMSM is a 1.65 hp. Unimotor UM series from Emerson. **Table 1** summarizes the technical specifications of the motor [27].

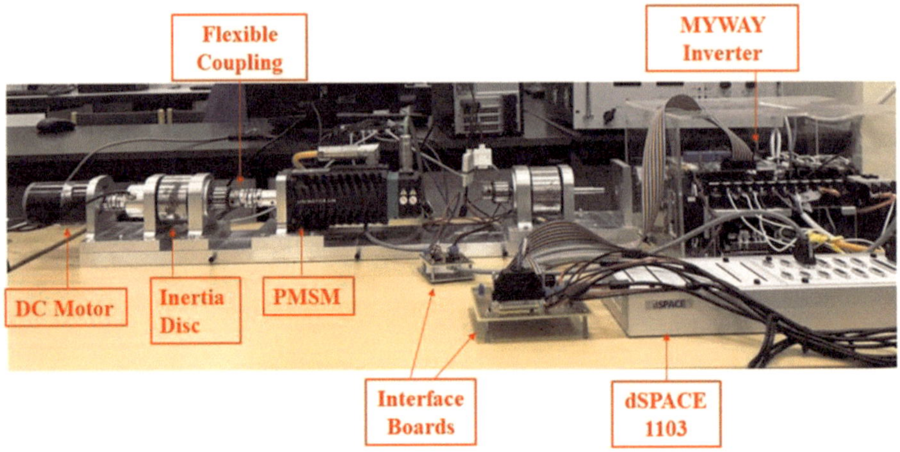

Figure 5.
PMSM experimental setup.

Variable	Value	Unit
Torque Constant (K_t)	1.6	Nm/A
EMF Coefficient (K_e)	98	V/krpm
Rated Torque (τ_{rated})	3.9	Nm
Stall Current (I)	2.7	A
Rated Speed (ω_{rated})	3000	rpm
Rated Power (P_{rated})	1.23	kW
Stator Resistance (R_{ph-ph})	6.8	Ω
Stator Inductance (L_{ph-ph})	24.3	mH
Inertia (J)	2.5	$kgcm^2$
Number of Poles (P)	6	—

Table 1.
PMSM parameters.

5.2 Initial motor alignment

For proper field orientation, an absolute encoder or a resolver is usually used with the PMSM. Therefore, an initial motor alignment is needed to operate the PMSM with the Sin/Cos encoder [28]. The main idea is to align the stator current vector is with the rotor flux vector Ψ_r along the stator phase a-axis. Hence, the angle between both vectors will be 0. This can be done by connecting phase a of the PMSM to the positive terminal of the DC link and connecting phases b and c to the negative terminal of the DC link. The application of a constant DC voltage will create a constant current flowing in all phases, and will place the stator current vector along the a-axis. This will in turn force the rotor flux vector to align itself with the stator current vector. This means that the d-axis is aligned with the a-axis, and the initial electrical angle of the rotor is 0. By this, the readings of an incremental encoder become sufficient to apply field orientation.

5.3 Synergetic controller testing

In this section, the performance of the SC with the modified macro-variable ψ_{1m}, proposed in Section 4-B, is compared to that of the conventional macro-variable. The performance of both macro-variables is tested by analyzing the behavior of id(t) andψ 1(t). Since the q-axis macro-variable ψ2(t) is not altered, the speed response and iq(t) remain unchanged. In this experiment, the PMSM is subjected to a speed reference step of 500 RPM followed by another step to 1000 RPM. The DC motor load is controlled to apply a load torque of 0:6 N.m. The load torque is then removed, and the speed reference is brought back to 0.

Figure 6 shows the speed reference profile and the speed response of both controllers. **Figure** 7 shows the response of the macro-variable $\psi_1(t)$ resulting from both controllers. The conventional macro-variable is given the subscript A and the proposed modified macro-variable is given the subscript B. The macro-variables and d-axis current signals have been filtered using a low pass filter with a cutoff frequency at 20 Hz to reduce noise and signal ripples.

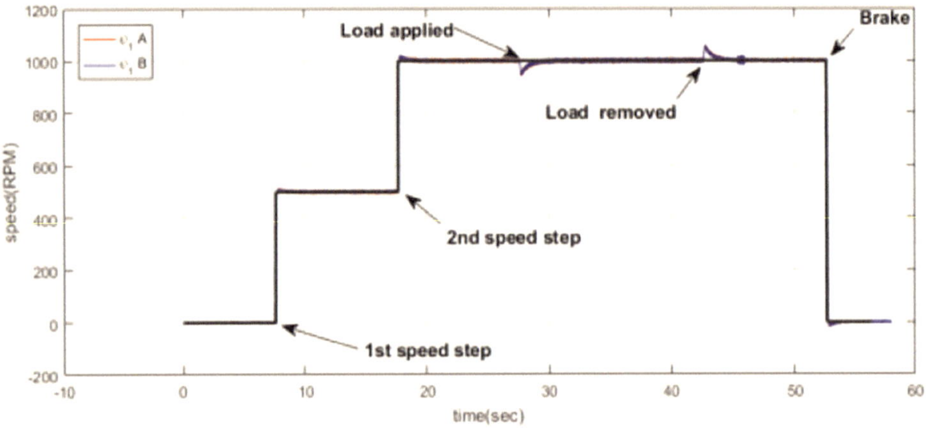

Figure 6.
Speed reference profile.

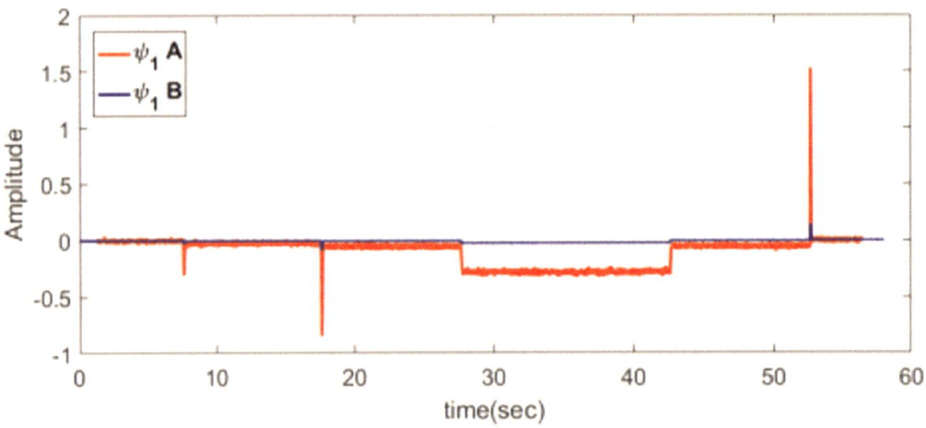

Figure 7.
Macro-variable ψ1 (t) response: (a) refers to the conventional macro-variable and (b) refers to the proposed macro-variable.

Figure 8 shows a zoomed view of each transient response. It can be seen that the proposed controller improves the response of $\psi_1(t)$ significantly. First, the macro-variable always converges back to zero after any transient event. The original controller, on the other hand, maintains a steady-state error and does not converge to zero as designed. Second, with the proposed controller, the magnitudes of the peak values reached during the transients are clearly much smaller than those reached by the original controller.

Table 2 summarizes the recorded peaks in **Figure 8**. It is shown that the proposed controller B improves the peak response of $\psi 1(t)$ by approximately 90%. The proposed controller shows good robustness to load changes. However, there still remains a very small steady state error in the response ofψ 1 m(t). The reason is that the internal friction torque and external load torque were not accounted for in the controller synthesis. Hence, all resistive load torques act as a disturbance to the system. It is worth noting that the response of controller B can be further improved by tuning the gains (K_4, K_5) in Eq. (16). The same previous analysis is done for the d-axis current

Transient Event	ψ_1 Max. Amplitude	ψ_{1m} Max. Amplitude	Improvement (%)
1st step reference	−0.3077	−0.02526	91.79
2nd step reference	−0.8425	−0.08622	89.77
Apply/Remove load	−0.2829	−0.02407	91.40
Braking	1.526	0.142	90.69

Table 2.
Summary results.

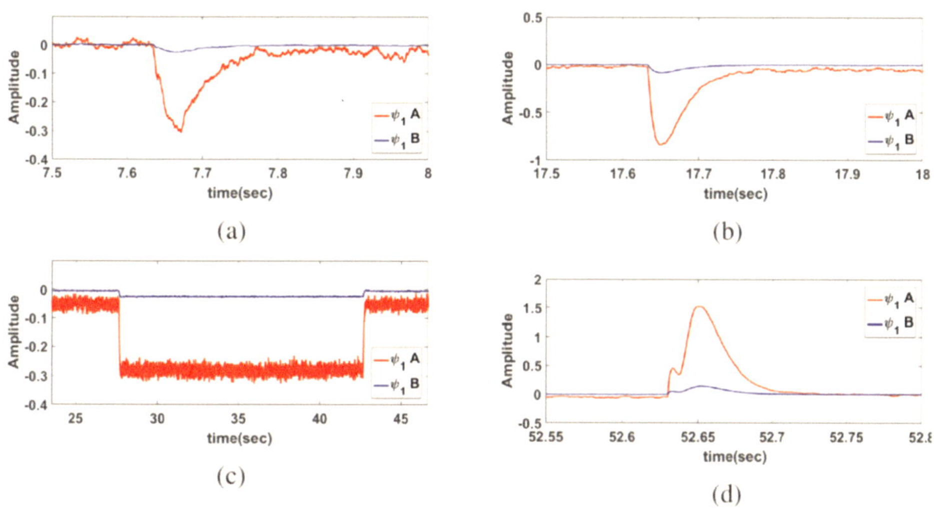

Figure 8.
Zoomed view at each transient event of ψ1(t): (a) 1st speed step (b) 2nd speed step (c) applying and removing load (d) braking.

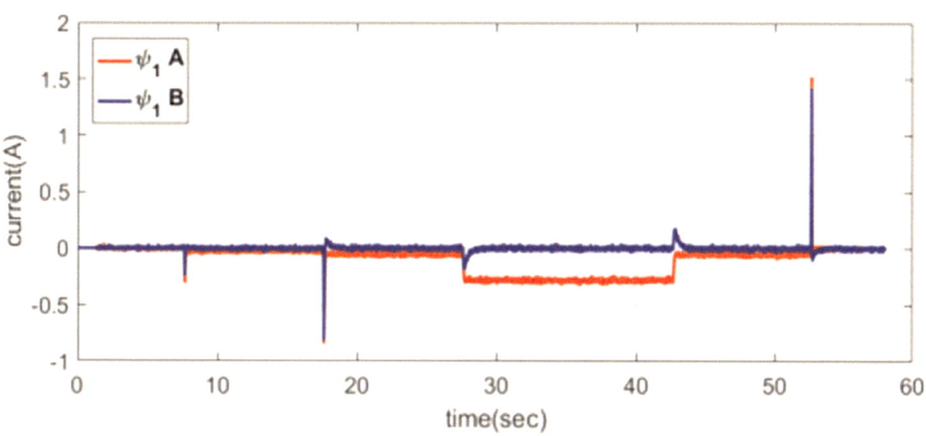

Figure 9.
Direct axis current id response.

$i_d(t)$. **Figure 9** shows the response of $i_d(t)$ for the two controllers (conventional vs. proposed) and **Figure 10** shows a zoomed view at each transient event. Studying the results, it can be seen that $i_d(t)$ exhibits lower peak magnitudes in the case of the

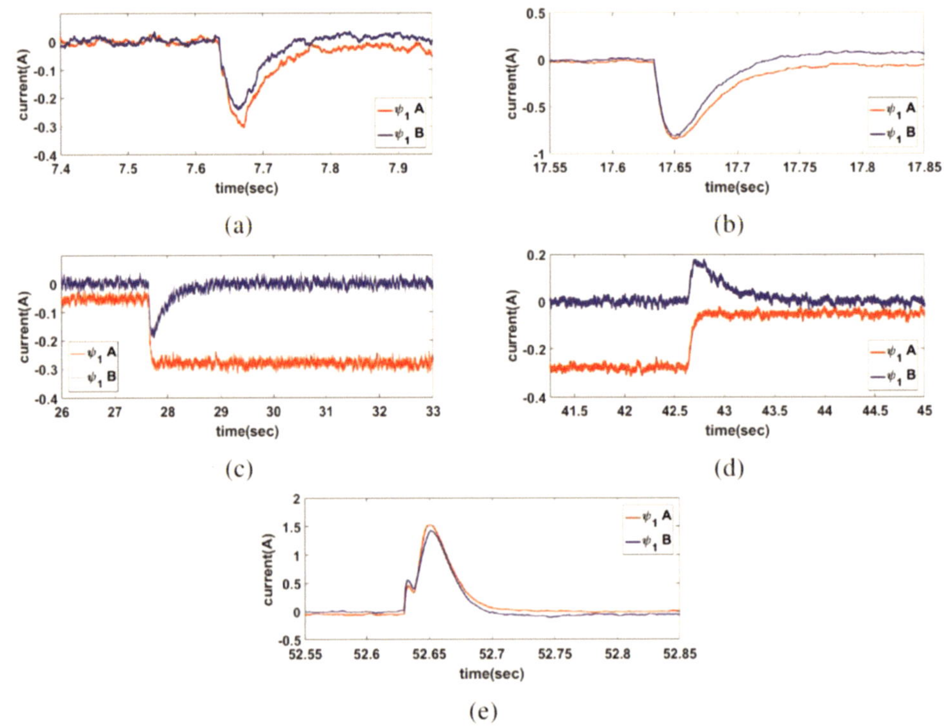

Figure 10.
Zoomed views at each transient period of i_d (t): (a) 1st speed step (b) 2nd speed step (c) applying load (d) removing load (e) braking.

proposed controller in all scenarios. Additionally, the proposed controller displays a very good disturbance rejection response. The current $i_d(t)$ maintains a zero- steady state error upon adding or removing a load torque.

5.4 Synergetic controller parameters tuning

In this section, the effect of all gains used in the proposed SC synthesis is studied. The controller tuning process is employed to find the best possible values for optimal performance. The same experimental procedure explained in Section 5.C is used to test the performance of the SC. Each gain is tuned while keeping the remaining gains constant to see the effect of this inspected gain separately. This procedure is done for each of the five gains $(K_1 - K_5)$ used in the controller design Eqs. (16)–(17) as well as the time constants T_d and T_q.

5.4.1 D-axis macro-variable ψ_1

The designed macro-variable ψ_1 uses two gains, K_1 and K_2, as shown in Eq. (16) and its dynamics depend on the time constant T_d. The effect of each parameter on the behavior of the macro-variable ψ_1 and the response of the d-axis current i_d will be analyzed. **Table 3** shows the selected gains to test the effect of K_1.

Figure 11 shows the response of the macro-variable ψ_1 upon changing the value of K_1 using the same speed reference profile as shown in **Figure 6**. Looking at the

K_1	0.01	0.05	0.1	0.15	0.2	0.25
K_2			0.3			
T_d			0.001 sec			

Table 3.
Summary of the selected values for K_1 inspection.

transient periods displayed in **Figure 12**, it can be observed that increasing the value of K_1 increases the magnitude of the peak value that ψ_1 reaches before it starts settling back to its reference. It is also noticed from **Figure 12c** that the steady state error when adding a load torque decreases for lower values of K_1. In the same line, it is noticed that increasing the value of K_1 increases the ripple of the macro-variable ψ_1.

Figure 13 shows the response of i_d for different values of K_1. It can be realized that i_d always converges to its reference value ($i_d = 0$) in steady state.

The same procedure is repeated to study the effect of K_2 on ψ_1 and i_d. **Table 4** summarizes the selected values of K_2 in this experiment.

It was observed that the gain K_2 does not affect the behavior of ψ_1 significantly. However, this is not the case for i_d. Changing the value of K_2 clearly affects the settling time of i_d especially when applying or removing a load as shown in **Figure 14**. Increasing K_2 results in a faster response.

It should be noted that all signals for ψ_1 and i_d are filtered by a low pass filter with a cut-off frequency of 20 Hz. It is also worth mentioning that these gains affect only the behavior of ψ_1 and i_d. The speed of the motor is not affected by changing these gains, since the speed and torque of the motor are controlled only by the second macro-variable ψ_2, which was not altered in this experiment.

Lastly, the effect of changing T_d on the behavior of ψ_1 and i_d is studied. **Table 5** shows the selected gains in this experiment. T_d is responsible for the time taken by ψ_1 to reach the manifold $\psi = 0$.

Figure 15 shows the response of ψ_1 upon changing the constant T_d. Changing T_d affects both the magnitude of the peaks as well as the steady state errors.

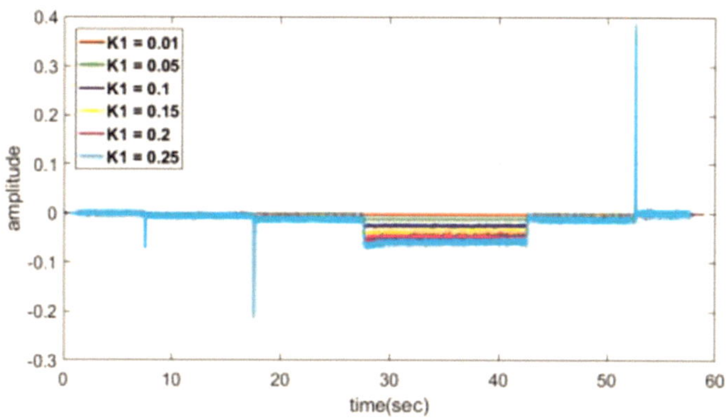

Figure 11.
Macro-variable ψ_1 response to different values of K_1.

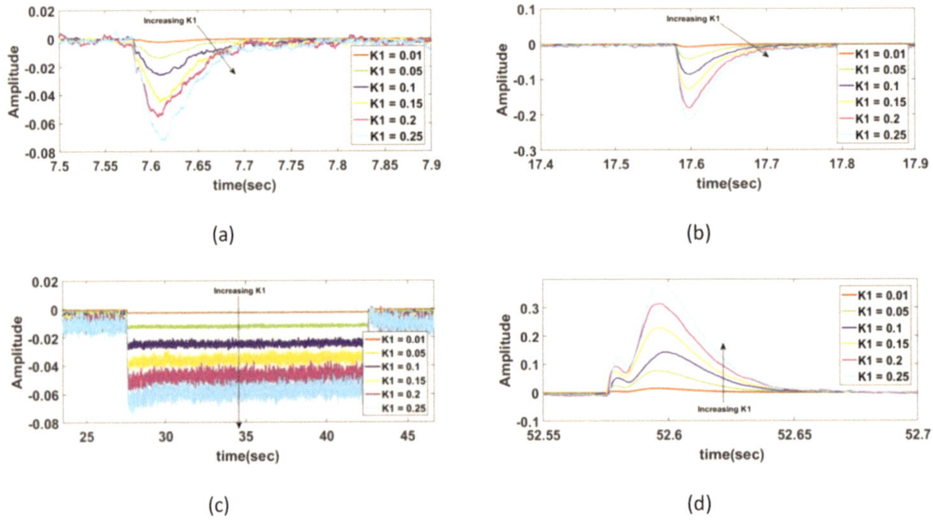

Figure 12.
Macro-variable ψ_1 response to different values of K_1. Zoomed views of each transient period: (a) 1st speed step, (b) 2nd speed step, (c) applying and removing load, (d) braking.

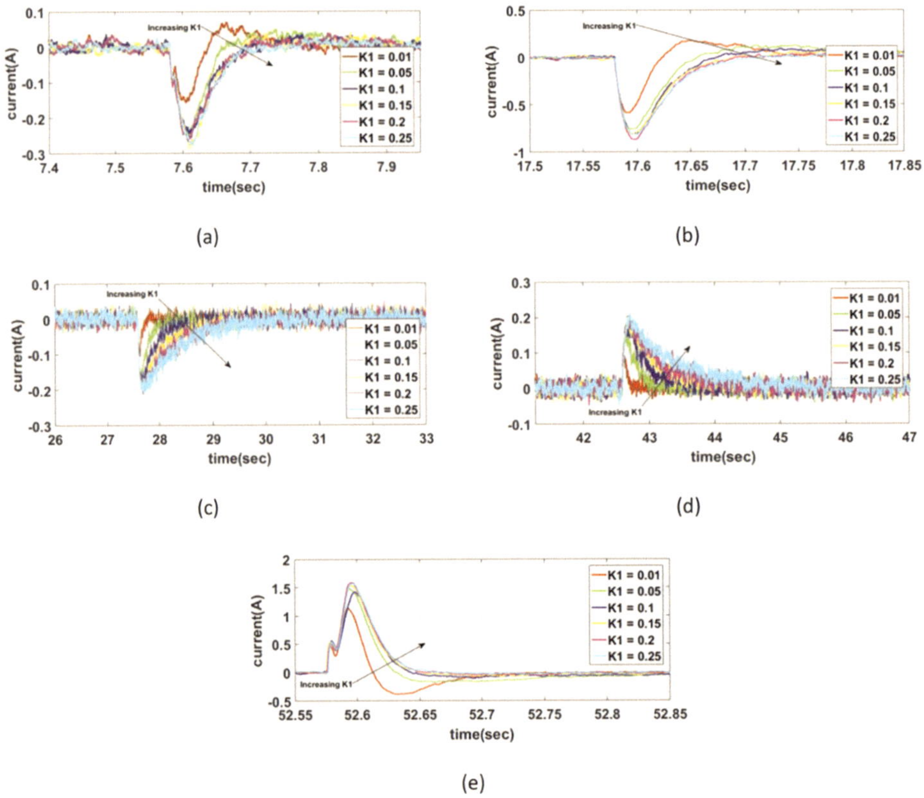

Figure 13.
D-axis current i_dresponse to different values of K_1. Zoomed views of each transient period: (a) 1st speed step, (b) 2nd speed step, (c) applying load, (d) removing load, (e) braking.

K_1			0.1			
K_2	0.05	0.1	0.15	0.2	0.25	0.3
T_d			0.001 sec			

Table 4.
Summary of the selected values for K_2 inspection.

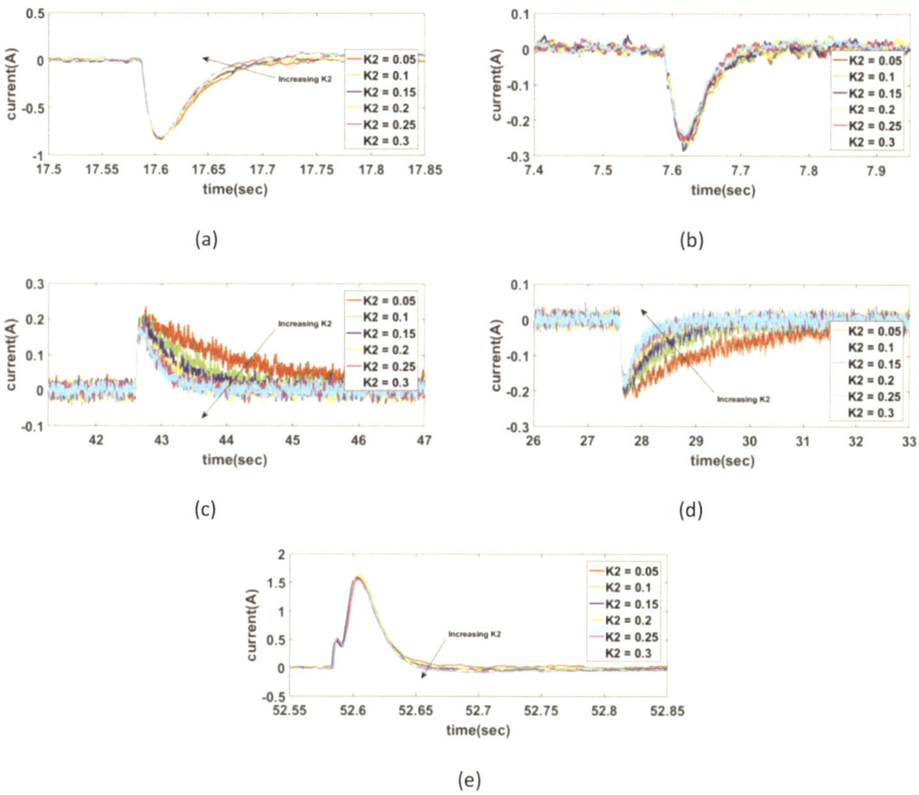

Figure 14.
D-axis current i_dresponse to different values of K_2. Zoomed views of each transient period: (a) 1st speed step, (b) 2nd speed step, (c) applying load, (d) removing load, (e) braking.

K_1		0.1	
K_2		0.3	
T_d	0.003	0.001	0.0007

Table 5.
Summary of the selected values for T_d inspection.

5.4.2 Q-axis macro-variable ψ_2

The designed macro-variable ψ_2 uses three gains, $K_3, K_4,$ and K_5, as shown in Eq. (17) and its dynamics depend on the time constant T_q. The effect of each parameter on the behavior of the macro-variable ψ_2 and the response of the q-axis current i_q was analyzed. K_3 and K_5 directly affect the speed, while K_4 affects the current i_q,

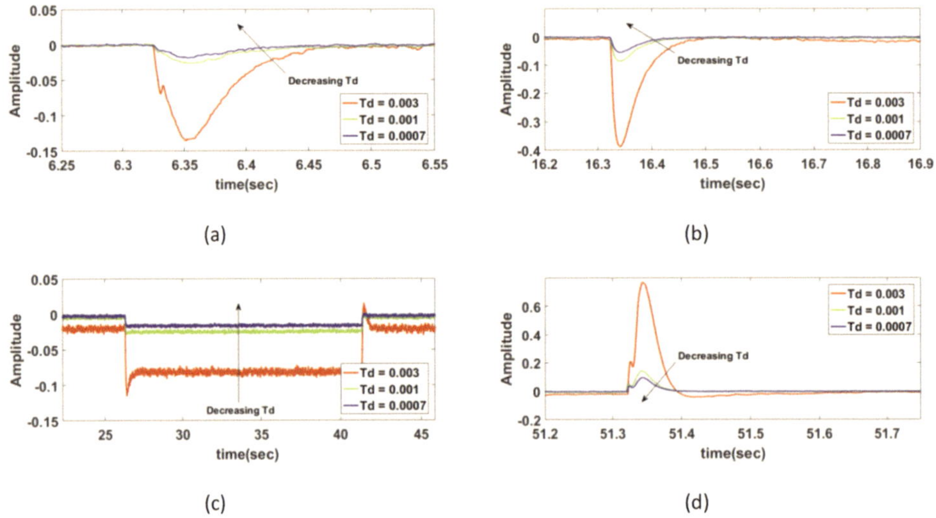

Figure 15.
Macro-variable ψ_1response to different values of T_d. zoomed views of each transient period: (a) 1st speed step, (b) 2nd speed step, (c) applying and removing load, (d) braking.

K_3	0.01	0.05	0.1	0.15	0.2
K_4			1		
K_5			0.15		
T_q			0.001 sec		

Table 6.
Summary of the selected values for K_3 inspection.

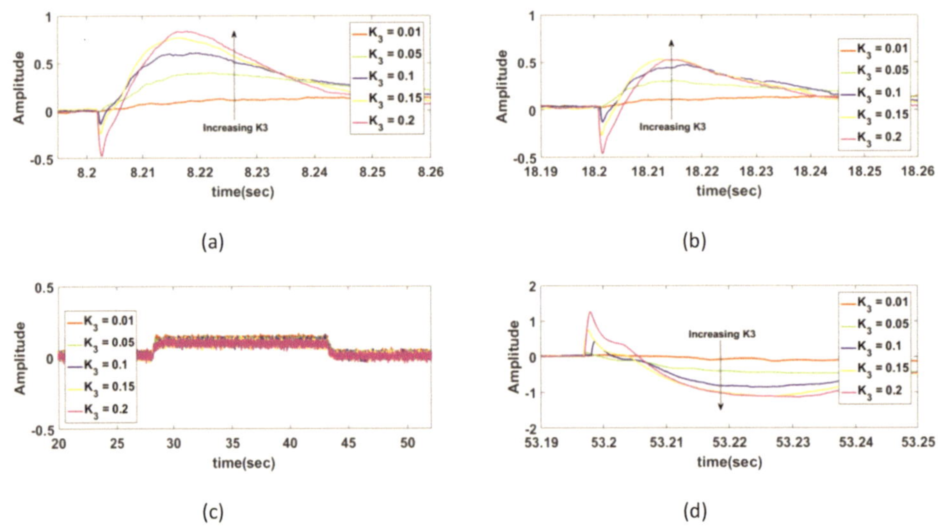

Figure 16.
Macro-variable ψ_2response to different values of K_3. Zoomed views of each transient period: (a) 1st speed step, (b) 2nd speed step, (c) applying and removing load, (d) braking.

which adjusts the machine torque. **Table 6** shows the selected gains to test the effect of K_3.

Figures 16 and **17** show the response of the motor speed and the micro-variable ψ_2 to different values of K_3.

It is observed that the amplitude of ψ_2 increases when K_3 is increased. The motor speed shows an underdamped response for transient for low values of K_3. And an overdamped behavior for large values of K_3. Similar analysis is made to study the effect of the gain K_4 and K_5. it is noticed that increasing K_5 increases the speed of convergence and eliminates steady state error faster. On the other hand, decreasing K_4 increases the speed of convergence of the motor speed and ψ_2.

Finally, the value of the constant T_q is changed to investigate its effect on the convergence speed of ψ_2. **Table 7** shows the selected gains in this experiment.

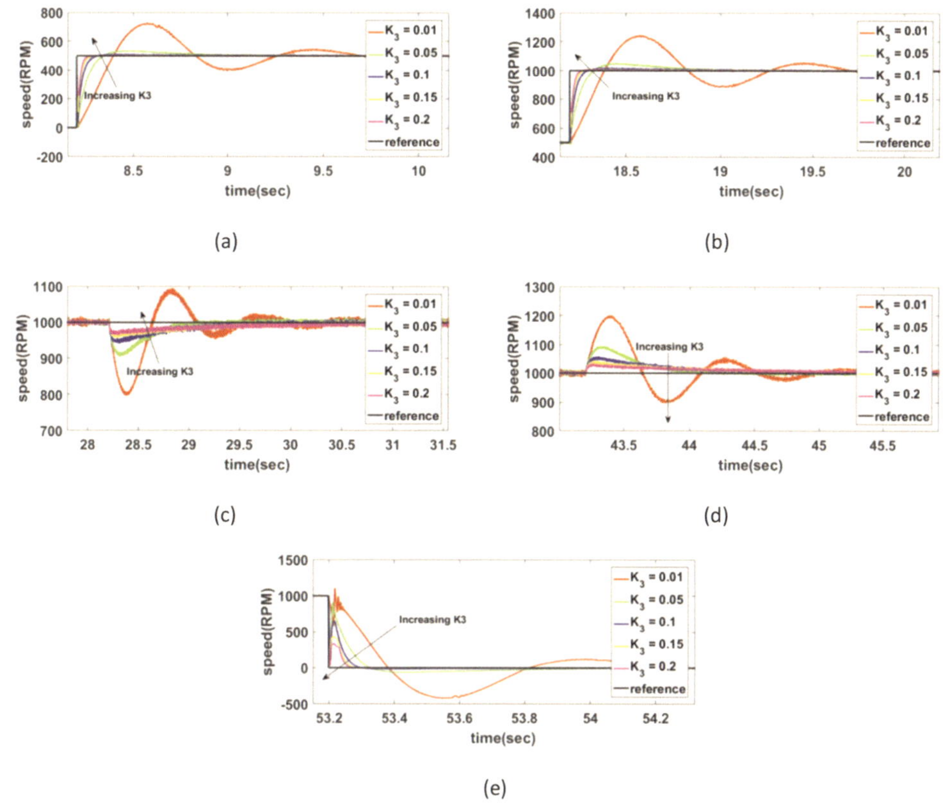

Figure 17.
Motor speed response to different values of K_3. Zoomed views of each transient period: (a) 1st speed step, (b) 2nd speed step, (c) applying load, (d) removing load, (e) braking.

K_3		**0.1**	
K_4		1	
K_5		0.15	
T_q	0.003	0.001	0.0007

Table 7.
Summary of the selected values for T_q inspection.

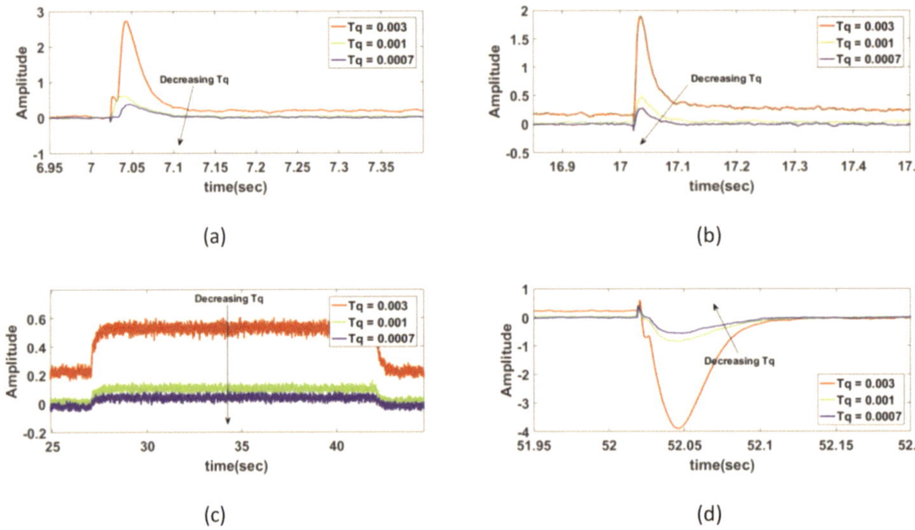

Figure 18.
Macro-variable ψ_2response to different values of T_q. zoomed views of each transient period: (a) 1st speed step, (b) 2nd speed step, (c) applying and removing load, (d) braking.

Figure 18 shows the response of ψ_1 upon changing the constant T_d. Changing T_d affects both the magnitude of the peaks as well as the steady state errors.

6. Comparison between synergetic control and field oriented control

6.1 Speed reference tracking and disturbance rejection

In this section, the performance of the proposed synergetic controller (SC) is compared to that of the field-oriented controller (FOC), which is used as a benchmark.

Both controllers are tuned so that their responses have approximately similar settling times, around 0.05 s, and are tested under the same operating conditions. The following characteristics are used for comparison:

- Settling time and steady-state error in response to a step reference speed and to a load step disturbance.
- Transient response of the d-axis current.
- Current harmonics.
- Response to wide speed variation.

Figure 19 shows the speed profile used, which consists of a sequence of speed step references following the application and removal of a load step disturbance.

The observed results indicate that both controllers track the speed reference as desired without any steady-state errors. Both controllers are able to correct the speed after adding or removing the load. However, it was observed that when starting from rest, the SC shows a slightly faster response with a settling time of 0.076 s as opposed to

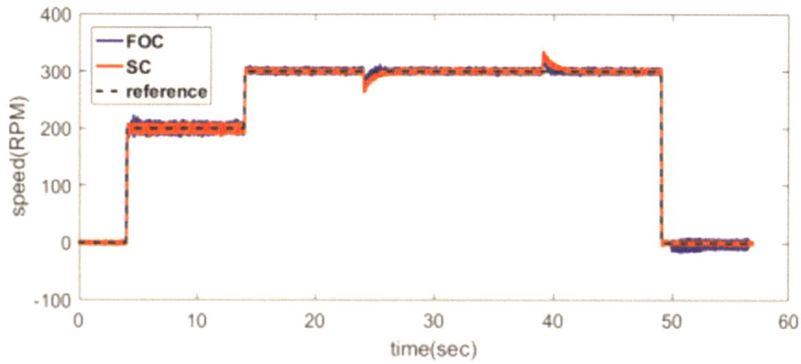

Figure 19.
Speed response of FOC and SC.

0.117 s for FOC. This shows that the speed of convergence, in the case of the SC, is better than that of the FOC by 35.042%, indicating that the SC was able to overcome the static friction that resists the motion of the motor when at rest faster than the FOC.

When both controllers are running at a fixed speed and the reference is increased, both exhibit the same settling time. Static friction acts as a disturbance to the system and the SC is able to overcome this disturbance more efficiently. This observation was confirmed by repeatedly conducting the step response experiment starting from rest for various speed step reference amplitudes. **Figure 20** illustrates sample results obtained with step reference values of 50 rpm and 200 rpm. We can therefore conclude from this analysis that the SC has better performance during motor startup under low-speed references.

Looking next at the d-axis current corresponding to the speed profile given in **Figure 19**, the following observations are made. Both controllers are able to maintain $i_d(t)$ at 0 without any steady state errors. However, $i_d(t)$ with SC shows a faster response, less current ripple in steady state, as well as lower peaks values (seen in **Figure 21** for the 1st step reference and the 1st load application). The d-axis i_d current deviation for FOC could be due to friction and sensitivity to system parameters. Overall, it can be concluded that FOC has a good dynamic and steady state characteristics however, SC shows better performance.

When looking at the three phase stator currents, the waveforms from both controllers are sinusoidal. However, the stator current waveforms by the FOC controller are slightly distorted compared to the smoother waveforms obtained by the SC. This means that SC results in less harmonics in the output waveforms. **Figure 22** shows the steady

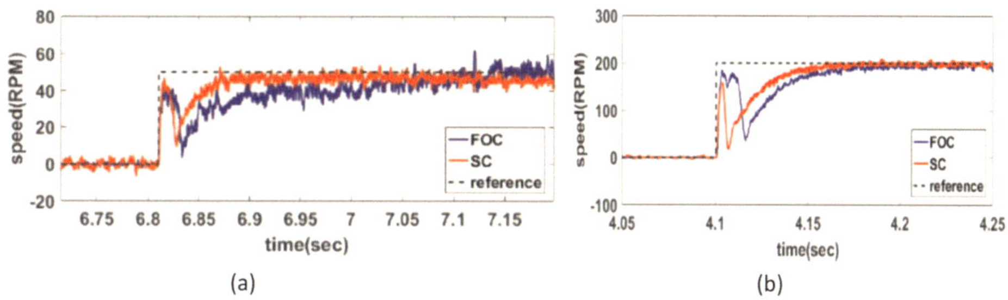

Figure 20.
Zoomed views of the motor speed response starting from rest for a reference speed step of (a) 50 rpm, and (b) 200 rpm.

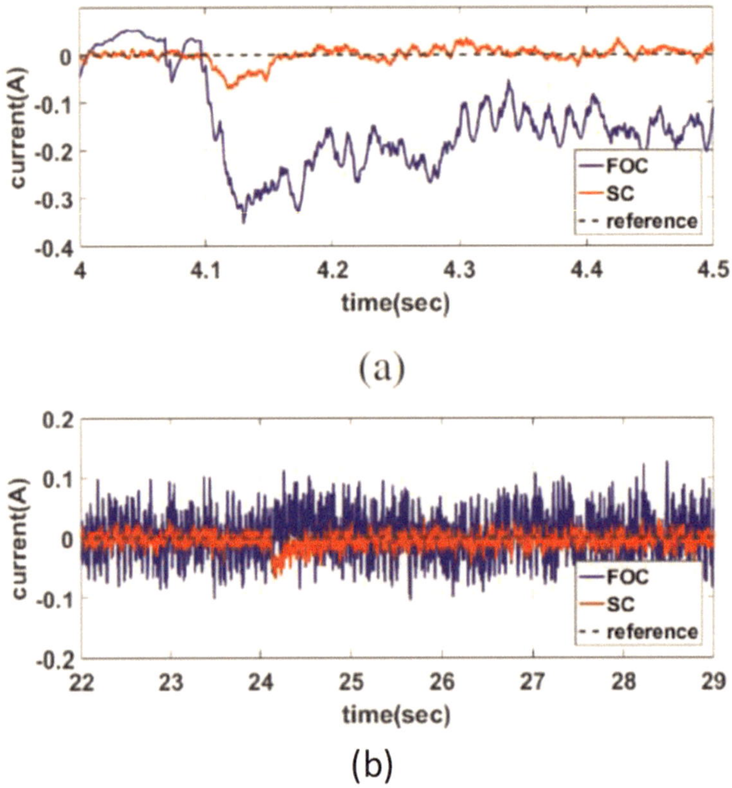

Figure 21.
Zoomed views at the transient response of id(t)(a) 1st speed step (b) applying load.

state currents waveforms at 300 rpm (15 Hz). The shown waveforms are filtered by means of a 16-sample moving averaging filter. The Fourier transform is next used to analyze the harmonics present in the current waveforms for each controller.

Figure 23 shows the single-sided amplitude spectrum of phase-a current for both controllers. Both controllers have the maximum magnitude at 15 Hz which is the

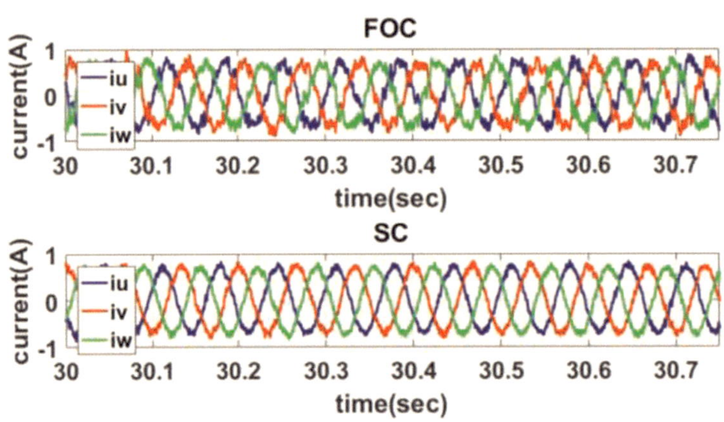

Figure 22.
Zoomed view of the three phase currents response at 300 rpm.

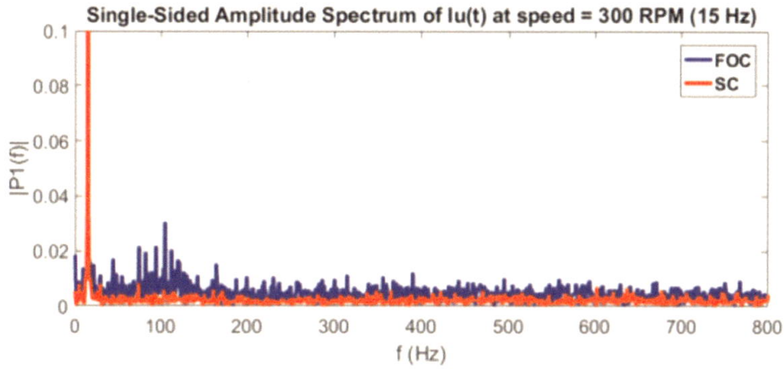

Figure 23.
Magnitude response of the FFT.

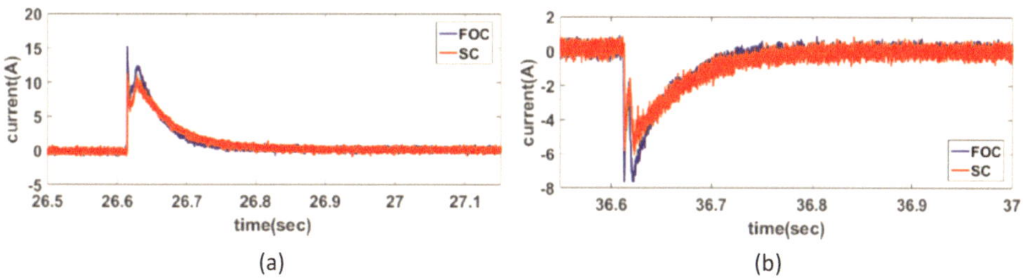

Figure 24.
Response of iq(t) to large speed variations (a) speed step from −1000 rpm to 1000 rpm (b) speed step from 1000 rpm to 0 rpm.

fundamental frequency. However, the FOC amplitude spectrum shows higher magnitudes at higher frequencies as opposed to the SC.

The last comparative analysis is for the performance of the controllers to wide speed variation. This is tested by applying a square wave speed reference ranging between −1000 rpm to 1000 rpm. The results show that both controllers are able to track the speed reference and reach zero steady state error. However, the SC reaches the reference faster. In addition, the SC shows less current amplitude during each speed transient (phase current amplitude is 26.4% lower for the second speed step) as shown in **Figure 24**. This makes the SC more robust in wide signal variations, as it requires less control effort to overcome the large speed variation.

6.2 Regenerative braking

In this section, regenerative braking under FOC and SC is implemented, and its performance is analyzed. Torque control is used to run the motor under RB mode.

The reference q-axis currents evaluated by Eq. (20) is used to generate the required electromagnetic torque and guarantee maximum recovered power. The maximum torque is given by

$$T_{reg-max} = -\frac{9P^2}{32R}\lambda_{PM}^2\omega \tag{24}$$

Figure 25 shows the topology of the voltage source inverter used to supply the motor. The DC bus voltage is rectified from a 380 V-rms 3-phase line. Additional filter capacitors are used to maintain the DC bus voltage at 575 V. In order to observe the voltage variation during regenerative braking, the three-phase line is disconnected from the inverter. As a result, the DC link voltage begins to drop while the motor is operating under a constant speed control mode.

Once the DC link voltage reaches 300 V, the regenerative braking command signal is triggered, and the motor operates under RB mode.

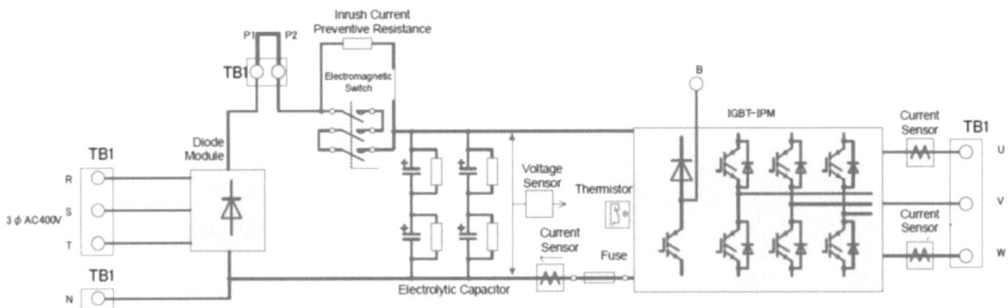

Figure 25.
Voltage source inverter supplying the PMSM.

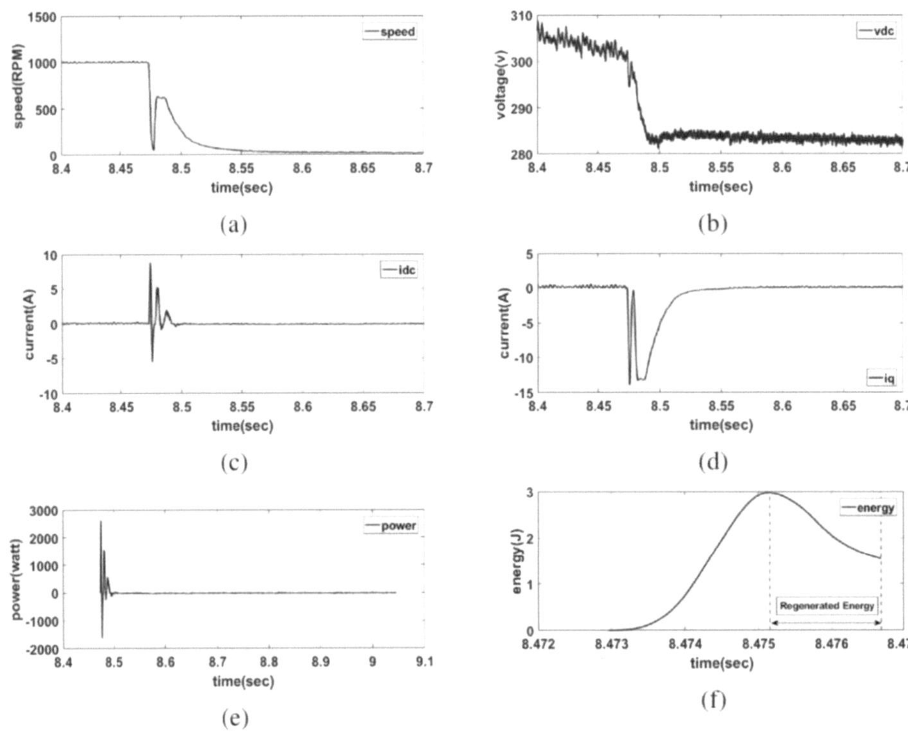

Figure 26.
Regenerative braking under torque control using FOC (a) motor speed (b) DC link voltage (c) DC link current (d) motor iq (e) DC link power (f) DC link energy.

The regenerated power is calculated by multiplying the DC link voltage by the DC link current during RB. The regenerated energy is next calculated by integrating the power by means of numerical integration following the trapezoidal rule.

$$P = V_{DC} I_{DC} \tag{25}$$

$$E_{reg}(k) = E_{reg}(k-1) + \frac{T_s}{2}[P(k) + P(k-1)] \tag{26}$$

The mechanical energy of the motor is given by

$$E_{mech} = \frac{1}{2} J \omega^2 \tag{27}$$

The efficiency of the regenerative braking process in terms of harvested energy is computed by

$$\eta = \frac{E_{reg}}{E_{mech}} \times 100. \tag{28}$$

Figure 26 shows the motor speed, DC link voltage, DC link current, q-axis current, regenerated power and energy on the DC link during the braking period under FOC. The motor comes to rest in a shorter period as well with a relatively high q-axis

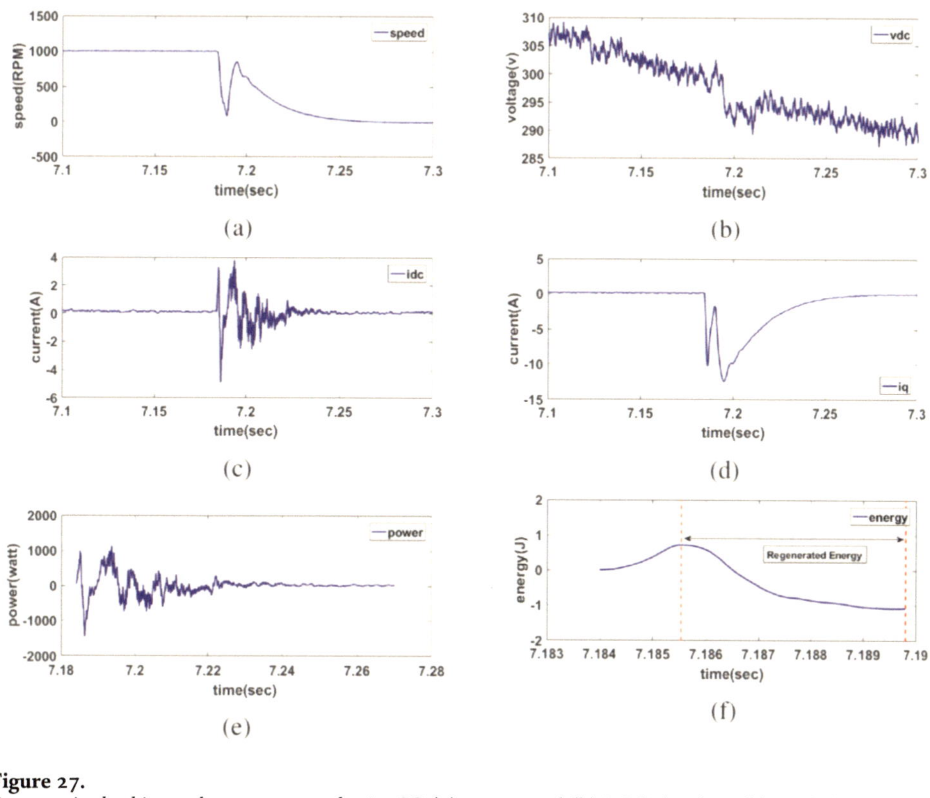

Figure 27.
Regenerative braking under torque control using SC (a) motor speed (b) DC link voltage (c) DC link current (d) motor iq (e) DC link power (f) DC link energy.

	i_{dc-max} (A)	ΔV_{dc} (V)	E_{reg} (J)	P_{max} (W)	i_{q-max} (A)	E_{mech} (J)
FOC	−5.485	5.6	1.450	1626	−13.97	17.272
SC	−4.893	6.0	1.801	1452	−12.44	17.272

Table 8.
Summary of the regenerative braking results under torque control.

current value. This indicates that the generated electromagnetic torque is high resulting in a fast braking.

The same experimental procedure is repeated under the SC. The macro-variables designed in section IV are used to control the motor in the braking period. **Figure 27** shows the inverter and motor variables during braking under SC. Comparing the waveforms resulting from both controllers, it can be concluded that the motor behaves more or less the same way under both controllers and the waveforms are highly similar. **Table 8** summarizes the results of the regenerative braking process under torque control using both controllers.

The oscillations in the motor speed caused by the flexible coupling hinder the controllers' performances, since the generated q-axis current $i_q(t)$ is a function of the motor speed. Therefore, the performance of the controllers in regenerative braking using this technique cannot be assessed properly unless a rigid coupling is used.

7. Conclusions

This chapter presented an optimum design of synergetic controllers for a permanent magnet synchronous motor (PMSM) drive system. We also provided an extensive study of the controller parameters tuning for optimal performance. The proposed macro-variables of the SC improve the performance of the synergetic controller significantly compared to the conventional controllers proposed in the literature by eliminating the steady-state errors in the direct axis current i_d. The results also show that the error in the macro-variables ψ_1 was improved by 90%. The synergetic controller showed also more robustness against disturbances, such as static friction, than the field-oriented controller. Upon starting from rest, conducted experiments show that the SC performs better than FOC in terms of convergence speed at low-speed reference values. The synergetic controller also reduces the harmonics of the system compared to the FOC using both the same SPWM modulator and switching frequency. Since one of the advantages of the SC is operating under low bandwidths, this was validated using the Fourier transform of the phase currents, where SC waveforms show fewer distortions. The results show that FOC adds more harmonics at multiples of the fundamental frequency of the motor. The voltage waveforms resulting from the SC are different from those resulting from FOC due to the nonlinear output command law. However, the fundamental component remains sinusoidal as desired. Both controllers have been tested in case of wide signal variations, the results show that the SC shows better performance against wide variations, where it is able to converge in shorter time while using lesser amounts of energy compared to FOC. This confirms the benefit of using a nonlinear controller in nonlinear systems, where the linear PI controllers used in FOC respond poorly to wide signal variations as opposed to the SC. Finally, the results indicate that regenerative braking using the torque control mode allows for energy harvesting. The SC displays a similar energy harvesting efficiency compared to the FOC. However, the oscillations in the motor speed caused by the flexible coupling hinder the controllers' performances, since the generated q-axis current i_q is a function of the motor speed.

Author details

Andrew Adib and Rached Dhaouadi*
College of Engineering, American University of Sharjah, Sharjah, UAE

*Address all correspondence to: rdhaouadi@aus.edu

References

[1] Amezquita-Brooks L, Liceaga-Castro J, Liceaga-Castro E. Speed and position controllers using indirect field-oriented control: A classical control approach. IEEE Transactions on Industrial Electronics. 2014;**61**(4): 1928-1943

[2] Yano M, Abe S, Ohno E. History of power electronics for motor drives in Japan. In: IEEE Conference on the History of Electronics. Bletchley Park, UK: Bletchley Park Trust; 2004. pp. 28-30

[3] Wang F, Zhang Z, Mei X, Rodríguez J, Kennel R. Advanced control strategies of induction machine: Field oriented control, direct torque control and model predictive control. Energies. 2018;**11**(120):1-13

[4] Szczepanski R, Tarczewski T, Grzesiak LM. PMSM drive with adaptive state feedback speed controller. Bulletin of the Polish Academy of Sciences: Technical Sciences. 2020;**68**(5): 1009-1017

[5] Łebkowski, "A way of neodymium-iron-boron magnets regeneration in surface-mounted PMSM used in electric vehicles," Bulletin of the Polish Academy of Sciences: Technical Sciences, Vol. 65, No. 5, pp. 751-758, 2017.

[6] Janaszek M. Structures of vector control of n-phase motor drives based on generalized Clarke transformation. Bulletin of the Polish Academy of Sciences: Technical Sciences. 2016; **64**(4):865-872

[7] Murthy S. Analysis of Regenerative Braking in Electric Machines (Thesis). Atlanta, GA, USA: Georgia Institute of Technology; 2013. Available from: http://hdl.handle.net/1853/47660

[8] Naseri F, Farjah E, Ghanbari T. An efficient regenerative braking system based on battery/supercapacitor for electric, hybrid, and plug-in hybrid electric vehicles with BLDC motor. IEEE Transactions on Vehicular Technology. 2017;**66**(5):3724-3738

[9] Chen C-H, Chi W-C, Cheng M-Y. Regenerative braking control for light electric vehicles. In: Power Electronics and Drive Systems (PEDS), 2011 IEEE Ninth International Conference on IEEE, Singapore. 2011. pp. 631-636. DOI: 10.1109/PEDS.2011.6147317

[10] El Ouanjli N, Derouich A, El Ghzizal A, Motahhir S, Chebabhi A, El Mourabit Y, et al. Modern improvement techniques of direct torque control for induction motor drives: A review. Protection and Control of Modern Power Systems. 2019;**4**(11):1-12

[11] Salem FB, editor. Direct Torque Control Strategies of Electrical Machines. London, United Kingdom: IntechOpen; 2021 [Online]. Available from: https://www.intechopen.com/books/8686. DOI: 10.5772/intechopen.80103

[12] Pratapgiri S. Direct instantaneous torque controlled switched reluctance motor drive for fan type load and constant torque load. In: Switched Reluctance Motor - Concept, Control and Applications. London, United Kingdom: IntechOpen; 2017. Available from: https://www.intechopen.com/chapters/55678. DOI: 10.5772/intechopen.69280

[13] Dianguo Xu, Bo Wang, Guoqiang Zhang, Gaolin Wang, and Yong Yu, "A review of sensorless control methods for AC motor drives," CES Transactions on

Electrical Machines and Systems. March 2018;**2**(1):104-115

[14] Laribi M, Cheikh MSA, Larbes C, Essounbouli N, Hamzaoui A. A sliding mode and synergetic control approaches applied to induction motor. In: 3rd International Conference on Systems and Control, Algiers, Algeria. 2013. pp. 212-217. DOI: 10.1109/ICoSC.2013.6750861

[15] Islam MM, Siffat SA, Ahmad I, Liaquat M, Khan SA. Adaptive nonlinear control of unified model of fuel cell, battery, ultracapacitor and induction motor based hybrid electric vehicles. IEEE Access. 2021;**9**:57486-57509. DOI: 10.1109/ACCESS.2021.3072478

[16] Fu X, Li S. A novel neural network vector control technique for induction motor drive. IEEE Transactions on Energy Conversion. 2015;**30**(4):1428-1437

[17] Kivanc OC, Ozturk SB. Sensorless PMSM drive based on stator feedforward voltage estimation improved with MRAS multiparameter estimation. IEEE/ASME Transactions on Mechatronics. 2018; **23**(3):1326-1337

[18] Santi E, Monti A, Li D, Proddutur K, Dougal RA. Synergetic control for dc-dc boost converter: Implementation options. IEEE Transactions on Industry Applications. 2003;**39**(6):1803-1813

[19] Kolesnikov G, Veselov A, Kolesnikov A, Monti F, Ponci ES, Dougal R. Synergetic synthesis of DC-DC boost converter controllers: Theory and experimental analysis. In: APEC. Seventeenth Annual IEEE Applied Power Electronics Conference and Exposition (Cat. No.02CH37335), Dallas, TX, USA. Vol. 1. 2002. pp. 409-415. DOI: 10.1109/APEC.2002.989278

[20] Monti R, Dougal E, Santi DL, Proddutur K. Compensation for step load variations when applying synergetic control. In: Eighteenth Annual IEEE Applied Power Electronics Conference and Exposition, APEC'03, Miami Beach, FL, USA. Vol. 1. 2003. pp. 334-340. DOI: 10.1109/APEC.2003.1179235

[21] Bastos J, Monti A, Santi E. Design and implementation of a nonlinear speed control for a PM synchronous motor using the synergetic approach to control theory. In: 2004 IEEE 35th Annual Power Electronics Specialists Conference (IEEE Cat. No.04CH37551), Aachen, Germany. Vol. 5. 2004. pp. 3397-3402. DOI: 10.1109/PESC.2004.1355075

[22] Bogani T, Lidozzi A, Solero L, Napoli AD. Synergetic control of PMSM drives for high dynamic applications. In: 2005 IEEE International Conference on Electric Machines and Drives, San Antonio, TX, USA. 2005. pp. 710-717. DOI: 10.1109/IEMDC.2005.19580

[23] Noei R, Kholerdi HA. Robust speed control of induction machine using synergetic controller. In: 2015 2nd International Conference on Knowledge-Based Engineering and Innovation (KBEI), Tehran, Iran. 2015. pp. 880-885. DOI: 10.1109/KBEI.2015.7436159

[24] Niu MZ, Wang T, Zhang Q, He X, Zhao ML. A new speed control method of induction motor. In: 35th Chinese Control Conference (CCC 2016), Chengdu, China. 2016. pp. 10140-10143. DOI: 10.1109/ChiCC.2016.7554961

[25] Santi E, Monti A, Li D, Proddutur K, Dougal R. Synergetic control for power electronics applications: A comparison with the sliding mode approach. Journal of Circuits, Systems, and Computers. 2004;**13**(04):737-760

[26] Inverter Unit MWINV-9R144 Hardware User's Manual. Vol. 4, rev. 1. Japan: MYWAY Plus Co.; 2012. Available

from: https://global.myway.co.jp/catalog.html

[27] Unitmotor UM. Product datasheet. Emerson Industrial Automation. CT Dynamics. 2003. Available from: https://www.artisantg.com/PLC/65285-1/Emerson-Control-Techniques-190UMC300CACAA-Unimotor-UM-Brushless-Servo-Motor

[28] Slapak V, Kyslan K, Mejdr F, Durovsky F. Determination of initial commutation angle offset of permanent magnet synchronous machine-an overview and simulation. Acta Electrotechnica et Informatica. 2014; **14**(4):17-22

Chapter 6

Fault Tolerant Control of Five-Phase Induction Motor Drive

Md. Habibullah, Tusar Debanath and Md. Shahadath Hossain Sabbir

Abstract

The fault-tolerant capability of multiphase induction motor (IM) drives without adding extra hardware has been an interesting research subject in recent times. Regardless of the application and reliability requirements, fault tolerance is obtained by the software. Although different types of faults may occur, the most considered is the open-phase one which leads to a reduction in the number of active phases in the multiphase drive. Predictive current control (PCC) was recently proposed in the literature for managing the post-fault operation of the drives when an open-phase fault is considered. In PCC, the motor torque is controlled indirectly by controlling the motor current. Predictive torque control (PTC) can control the motor torque directly. However, PTC-based fault tolerant control of a five-phase IM (FPIM) drive has not been published in the literature. Hence, this fault-tolerant capability using the PTC method for an FPIM has been discussed in this chapter. Detail about the modeling of an FPIM, fault detection, and controller for both healthy and faulty conditions of IM has been discussed. The same model of the drive under both healthy and faulty conditions is considered. But the applied cost function is changed in a faulty condition.

Keywords: five-phase induction motor, open-phase fault, fault detection, fault tolerant control, predictive torque control

1. Introduction

Since the late 1990s, Multiphase motors are getting gradual popularity over their counterpart three-phase motors in various application fields, especially where critical safety is required, for example, propulsion applications, electric aircraft, electric vehicles, etc. [1]. Multiphase motors can drive the load smoothly, even if one or more phases of the motors are damaged. Among the multiphase motors, the five-phase induction motors (FPIMs) are widely used [2]. They have become a useful replacement for the three-phase motors because of higher reliability, higher power handling capacity without exceeding the current handling capacity of semiconductor switches and insulations, and better torque performance. Although multiphase drives are generally claimed to be 'fault-tolerant', this term is somewhat broad since many different types of faults may appear in an electrical drive, including both inverter and machine faults that may lead to short-circuit (phase, inverter switch, inter-turn) or open-circuit (inverter switch, phase or line) faults. Among the aforementioned possible faults, the probability of open-phase fault (OPF) in a drive system is high [2].

Due to the advancement of faster microprocessors, model predictive control (MPC) has received wide attention over existing control algorithms for industrial drives such as field-orientation control (FOC), direct torque control (DTC), and proportional-resonant (PR) control. MPC uses plant models and digital control platforms and allows system constraints and restrictions in a very intuitive way. In different applications of motor drives, MPC is found superior to DTC, FOC, and PR controllers for its increased flexibility and faster torque response [2, 3]. The MPC for FPIM drives has two variants: one is predictive torque control (PTC) and another one is predictive current control (PCC). Both types of MPC can effectively control the torque, flux, and thus speed of the induction motors. In PCC, the motor torque and flux are controlled indirectly by controlling the motor current. On the other hand, in PTC, the motor torque and flux are controlled directly and thus comparatively faster torque response is achieved [3]. A PCC has been recently proposed for managing the post-fault operation of the drives when an OPF is considered [4]. The faulty situation assumes zero stator current while freewheeling diodes can continue conducting in a non-controlled mode. An analysis is presented on the post-fault operation of the five-phase drive when the freewheeling diodes of the faulty phase are still conducting. Another PCC is presented in [5] for a six-phase IM drive and it is reported that PCC misbehaves in a post-fault situation if there is a significant delay in fault detection. The aforementioned PCC approaches need to be reconfigured when a fault occurs. A universal reconfiguration-less PCC approach is presented in [6] and it is shown that the system is naturally fault-tolerant. However, the fault-tolerant capability of an FPIM using PTC has not been stated yet in the literature. In this chapter, the performance of a PTC-based FPIM drive in both healthy and faulty conditions has been discussed.

2. System modeling

The block diagram of the proposed PTC for the FPIM is shown in **Figure 1**. The motor is fed by a two-level five-phase (2L–5Φ) voltage source inverter (VSI). The inverter is driven by the control signals which are generated by the PTC controller (a variant of MPC). The control system has four sub-sections: FPIM, 2L–5Φ inverter, PTC controller, and fault identification. The reference torque is generated by an outer speed loop and PI controller. PTC uses the measured stator current and dc-link voltage, and also receives the status of the machine whether it is healthy or faulty from the fault identification sub-section to predict the control objectives such as stator current (α–β), harmonic current (x–y), stator flux and torque. The predicted control objectives are compared with their corresponding references, and their error costs are calculated for all possible voltage vectors of a 2L–5Φ inverter. Based on the calculated error costs, the optimization function selects a switching state which yields minimum error cost and then applies the voltage vector corresponding to the selected switching state to the motor through the inverter. If OPF occurs, the fault identification section identifies the fault and signals the controller to reconfigure it. Because of the reconfigured control structure, the motor can run smoothly in a faulty condition with a de-rated torque capacity.

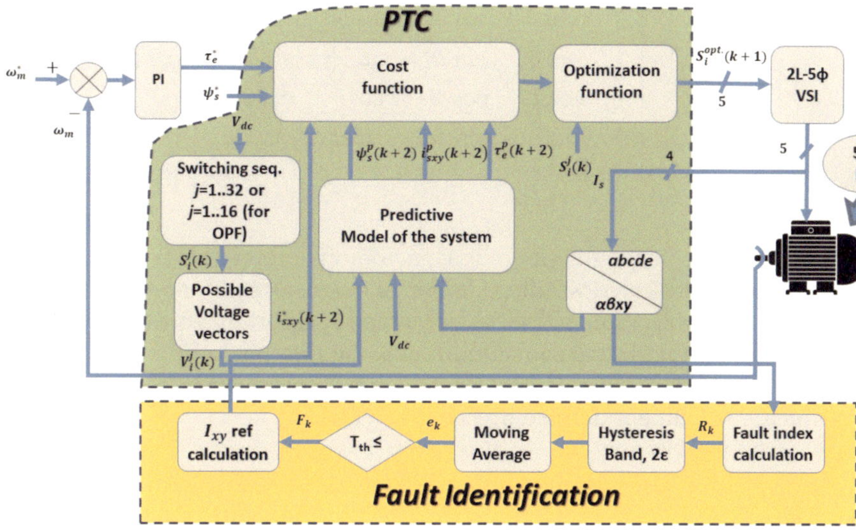

Figure 1.
Block diagram of the proposed fault-tolerant PTC of a five-phase induction motor drive.

2.1 Five-phase induction motor modeling

The FPIM with a squirrel cage rotor has distributed windings that are symmetrically displaced by $\theta = \frac{2\pi}{5}$ in between phases. The machine is powered by a 2L–5Φ VSI. As a general rule, the number of possible switching states N is,

$$N = A^B \tag{1}$$

Here, A is the number of possible states of each leg of the inverter and B is the number of phases of the inverter. So, for this instance, the number of switching states $N = 2^5 = 32$. This means there are 32 possible stator voltage vectors (30 active and 2 zero vectors). If S_i is denoted as the switching state of each inverter leg where $i = \{a, b, c, d, e\}$, the switching state is $[S_a \;\; S_b \;\; S_c \;\; S_d \;\; S_e]^T$ where $S_i \in \{0, 1\}$ designates that when the lower switch is on and the upper switch is off. The switching state and the dc link voltage V_{dc} may then be used to calculate the voltage for each stator phase using Eq. (2).

$$\begin{bmatrix} V_{as} \\ V_{bs} \\ V_{cs} \\ V_{ds} \\ V_{es} \end{bmatrix} = \frac{V_{dc}}{5} \begin{bmatrix} 4 & -1 & -1 & -1 & -1 \\ -1 & 4 & -1 & -1 & -1 \\ -1 & -1 & 4 & -1 & -1 \\ -1 & -1 & -1 & 4 & -1 \\ -1 & -1 & -1 & -1 & 4 \end{bmatrix} \begin{bmatrix} S_a \\ S_b \\ S_c \\ S_d \\ S_e \end{bmatrix} \tag{2}$$

Applying the current-invariant decoupling of Clarke's transformation, Eq. (2) can be mapped into two orthogonal stationary subspaces, $\alpha-\beta$ and $x-y$, plus a zero-sequence component, as shown in Eq. (3). Here, the stator voltage vectors have been identified using a decimal number equivalent to the binary number $[S_a \;\; S_b \;\; S_c \;\; S_d \;\; S_e]$.

$$\begin{bmatrix} V_{\alpha s} \\ V_{\beta s} \\ V_{xs} \\ V_{ys} \\ V_{zs} \end{bmatrix} = \frac{2}{5} \begin{bmatrix} 1 & \cos\theta & \cos 2\theta & \cos 3\theta & \cos 4\theta \\ 0 & \sin\theta & \sin 2\theta & \sin 3\theta & \sin 4\theta \\ 1 & \cos 2\theta & \cos 4\theta & \cos\theta & \cos 3\theta \\ 0 & \sin 2\theta & \sin 4\theta & \sin\theta & \sin 3\theta \\ \frac{1}{2} & \frac{1}{2} & \frac{1}{2} & \frac{1}{2} & \frac{1}{2} \end{bmatrix} \begin{bmatrix} V_{as} \\ V_{bs} \\ V_{cs} \\ V_{ds} \\ V_{es} \end{bmatrix} \quad (3)$$

Considering a sinusoidally distributed magnetomotive force (MMF), uniform air gap, symmetrically distributed windings, minimal magnetic saturation, and core losses, and with the application of a series of voltage equilibrium equations derived from the stator and rotor electromagnetic circuits, the five-phase IM may be represented in a stationary reference frame. Given this information, the functioning of the multiphase machine is defined by Eqs. (4)–(12).

$$V_s = I_s R_s + \frac{d\psi_s}{dt} \quad (4)$$

$$0 = I_r R_r + \frac{d\psi_r}{dt} - j\omega_e \psi_r \quad (5)$$

$$\psi_s = L_s I_s + L_m I_r \quad (6)$$

$$\psi_r = L_m I_s + L_r I_r \quad (7)$$

$$\tau_e = \frac{5}{2} p I_m \{\psi_r^* . I_s\} \quad (8)$$

$$j\frac{d\omega_m}{dt} = \tau_e - \tau_l \quad (9)$$

$$\omega_e = p\omega_m \quad (10)$$

$$\frac{dI_s}{dt} = \frac{1}{\tau_\sigma}\left[-I_s - \frac{K_r}{R_\sigma}\left(\frac{-\psi_r + j\tau_r\omega_e\psi_r}{\tau_r}\right) + \frac{1}{R_\sigma}V_s\right] \quad (11)$$

$$\frac{d\psi_r}{dt} = \frac{1}{\tau_r}(L_m I_s - \psi_r + j\omega_e \tau_r \psi_r) \quad (12)$$

The symbols or variables in Eqs. (4)–(12) have their usual meanings, as shown in **Table 1** given in Appendix. Integrating Eq. (9) with respect to time, the rotor angular speed ω_m can be derived and it is directly related to the rotor angular frequency ω_e by the pole pairs p, which is formulated in Eq. (10). Replacing the variable I_r with the variables ψ_r and I_s in Eq. (5) forms the modified rotor voltage in Eq. (12). Extracting I_r from Eq. (7) and putting its value in Eq. (6) develops the relationship between stator and rotor flux. Differentiating this stator flux with respect to time and placing it in the stator voltage equation emerges Eq. (11) in a compact form which aids in the prediction of stator current for the next sampling instant. To anticipate the torque, rotor flux, and stator current predictions are necessary. The compact factors used in IM modeling (Eqs. (11)–(12)) are given in Appendix (**Table 2**).

2.2 FPIM modeling under faulty condition

If a fault occurs in the machine, the controller detects the fault first. Then, the controller reconfigures itself to accommodate the fault and continues to operate the drive under faulty condition. Eq. (13) is verified when the motor drive is in healthy operation. But when an OPF occurs (assume the faulty phase is 'a'), the summation of healthy phase voltages is no longer equal to zero ($\sum V_{ks} \neq 0$, here $k \in \{b, c, d, e\}$). Then there comes the relation between α component and x component of stator current as in Eq. (15). The five vector space decomposition (VSD) [4] phase currents are no longer independent. Since phase voltage symmetry should remain valid in the post-fault situation, and the voltage of the faulty phase 'a' is no longer controllable due to its oscillation caused by the back EMF as in Eq. (14), the phase voltages must be modified accordingly to arrive at Eq. (16).

$$V_{as} + V_{bs} + V_{cs} + V_{ds} + V_{es} = 0 \tag{13}$$

$$V_{as} = I_{as}R_s + \frac{d\lambda_{as}}{dt} = \frac{d\lambda_{as}}{dt} = Back\ EMF_{\mathrm{a}}\ (\text{with } I_{as} = 0) \tag{14}$$

$$I_\alpha + I_x = 0 \Rightarrow \frac{-I_x}{I_\alpha} = 1 \tag{15}$$

$$\begin{bmatrix} V_{bs} \\ V_{cs} \\ V_{ds} \\ V_{es} \end{bmatrix} = \frac{V_{dc}}{4} \begin{bmatrix} 3 & -1 & -1 & -1 \\ -1 & 3 & -1 & -1 \\ -1 & -1 & 3 & -1 \\ -1 & -1 & -1 & 3 \end{bmatrix} \begin{bmatrix} S_b \\ S_c \\ S_d \\ S_e \end{bmatrix} - \frac{L_m \frac{dI_{as}}{dt} + L_m \frac{dI_{ar}}{dt}}{4} \cdot [I_4] \tag{16}$$

Where, $[I_4]$ is the identity matrix of order 4. The second term on the right-hand side of Eq. (14) is the counter EMF in terms of VSD variables. An asymmetrical system of equations depicting a two-phase machine with asymmetrical $d{-}q$ windings results from the VSD performed by the model proposed in [7] based on an orthogonal decoupling matrix. Since, there are four healthy phases, the number of available switching states is reduced from $2^5 = 32$ to $2^4 = 16$ in the faulty condition. So, to reduce the complexity, a non-orthogonal reduced-order transformation matrix is deduced from Eq. (3) which is shown in Eq. (17). The healthy motor model still can be used. The steady-state $\alpha{-}\beta$ current references are circular, not ellipsoidal in both pre-fault and post-fault operations.

$$\begin{bmatrix} V_{\alpha s} \\ V_{\beta s} \\ V_{ys} \\ V_{zs} \end{bmatrix} = \frac{2}{5} \begin{bmatrix} \cos\theta - 1 & \cos 2\theta - 1 & \cos 3\theta - 1 & \cos 4\theta - 1 \\ \sin\theta & \sin 2\theta & \sin 3\theta & \sin 4\theta \\ \sin 2\theta & \sin 4\theta & \sin 6\theta & \sin 8\theta \\ 1 & 1 & 1 & 1 \end{bmatrix} \begin{bmatrix} V_{bs} \\ V_{cs} \\ V_{ds} \\ V_{es} \end{bmatrix} \tag{17}$$

The new reduced-order Clarke transformation matrix permits the same set of $\alpha{-}\beta$ and $x{-}y$ equations in post-fault conditions as in healthy operating conditions. It should be noted that the controller is not informed of the inactive phase of the machine during the occurrence of the OPF. So, PTC still will regulate the machine with an incorrect model. The delay in fault identification has a more significant impact on PTC than FOC. That is why there should be an integrated fault detection technique in PTC as fast as possible.

Several pieces of literature have been published, focusing on different fault-tolerant techniques based on VSD variables (named VSDFD) [4, 8] and observing the phase currents [9]. Reference [8] analyzes the fault occurrence using fault indices that are directly dependent on $x{-}y$ current components. Fault identification using $x{-}y$

current components is simpler as their values are equal to zero while the machine is in healthy condition and these current components have no contribution to the production of flux or torque. The procedure for generating fault indices involves using the transformation matrix of Eq. (3) to produce the phase current equations and setting them equal to zero ($I_{ph} = 0$, the OPF condition). This results in the following set of equations:

$$R_a = -\frac{I_x}{I_\alpha} \tag{18}$$

$$R_b = \frac{I_x}{0.38I_\alpha + 1.17I_\beta - 0.73I_y} \tag{19}$$

$$R_c = \frac{I_x}{2.62I_\alpha - 1.90I_\beta + 3.08I_y} \tag{20}$$

$$R_d = \frac{I_x}{2.62I_\alpha + 1.90I_\beta - 3.08I_y} \tag{21}$$

$$R_e = \frac{I_x}{0.38I_\alpha - 1.17I_\beta - 0.7266I_y} \tag{22}$$

Here, $R_k (k \epsilon \{a, b, c, d, e\})$ is denoted as the fault index of different phases. In normal operating conditions, $R_k = 0$, whereas and in faulty condition, $R_k = 1$. That is why this method is particularly advantageous to detect and localize the fault. Since, phase '*a*' is considered as the faulty phase here, $R_a = -\frac{I_x}{I_\alpha} = 1$. But the controller must ensure that if there is an open phase fault apart from phase '*a*', it should consider the faulty phase as '*a*' and reorganize the other phases in sequential order. This action will allow the post-fault transformation matrix to ensure a smooth transition from pre-fault mode to post-fault mode and proper control performance. For instance, if we now consider phase '*b*' as the faulty phase, then R_b of Eq. (19) will be equal to 1 and this phase will now be marked as phase '*a*' and the other phases will be reorganized sequentially.

The ratios R_k can be integrated using the moving average method. Then a hysteresis band (2ε), as shown in **Figure 1**, is applied to the ratios of R_k. This hysteresis band ensures lower ripples (tends to zero) to the fault indices of healthy phases and generates filtered values of fault index ratio. These filtered values create new fault indices (e_k), as shown in **Figure 1**, during the period of the moving average. Finally, the fault indices are then compared to a threshold and determine the fault flag symbolized as $F_k \in \{0, 1\}$.

When any OPF occurs, the subspace orthogonal voltage components, α–β and x–y, must be calculated according to the modified transformation matrix as shown in Eq. (17). So, the following functions (Eqs. (23)–(27)) can be used to calculate the orthogonal subspace voltage components.

$$v_{\alpha s} = \frac{2}{5}((\cos\theta - 1)v_{bs} + (\cos 2\theta - 1)v_{cs} + (\cos 3\theta - 1)v_{ds} + (\cos 4\theta - 1)v_{es}) \tag{23}$$

$$v_{\beta s} = \frac{2}{5}(\sin\theta v_{bs} + \sin 2\theta v_{cs} + \sin 3\theta v_{ds} + \sin 4\theta v_{es}) \tag{24}$$

$$v_{xs} = -v_{\alpha s} \tag{25}$$

$$v_{ys} = \frac{2}{5}(\sin 2\theta v_{bs} + \sin 4\theta v_{cs} + \sin 6\theta v_{ds} + \sin 8\theta v_{es}) \tag{26}$$

2.3 Two-level five-phase inverter

The 2L–5Φ VSI has ten switches (two switches per leg), as shown in **Figure 2**. The number of total states is $2^5 = 32$ with which two zero vectors and thirty active vectors. The voltage vector expressions in $\alpha-\beta$ and $x-y$ planes can be written as,

$$v_{\alpha\beta} = v_\alpha + jv_\beta = 2/5\left(v_a + a\, v_b + a^2 v_c + a^3 v_d + +a^4 v_e\right) \tag{27}$$

$$v_{xy} = v_x + jv_y = 2/5\left(v_a + a^2 v_b + a^4 v_c + a^6 v_d + +a^8 v_e\right) \tag{28}$$

where, $a = e^{j\,2\pi/5}$ and $v_a \dots v_e$ are the phase voltages.
The relation between phase voltage v_x of phase x and switching states is,

$$v_x = S_x * V_{dc} \tag{29}$$

where the switching states of phase x are

$$S_x = \begin{cases} 1, \text{if the upper switch is on, the lower switch is off;} \\ 0, \text{ if the upper switch is off, the lower switch is on.} \end{cases}$$

The subspace voltage components produced by a 2L–5Φ VSI are plotted in $\alpha-\beta$ and $x-y$ planes, as shown in **Figure 3**.

The controller generates the optimum switching signal for the inverter as per the minimization of an objective function. The selected switching signal is then applied to the inverter. The inverter then supplies the required voltage to the induction motor.

2.4 PTC algorithm

The PTC works in three steps: the generation of the available voltage vectors for the inverter, the prediction of control objectives, and the selection of an optimal

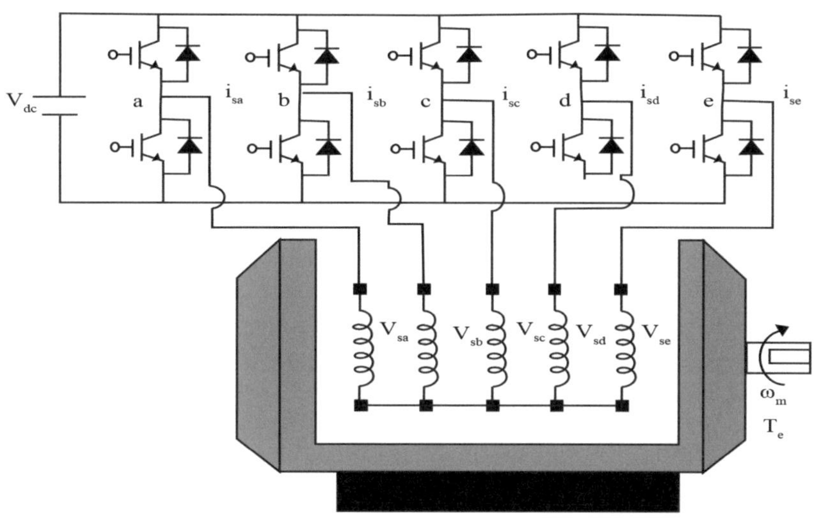

Figure 2.
FPIM drive schematic diagram.

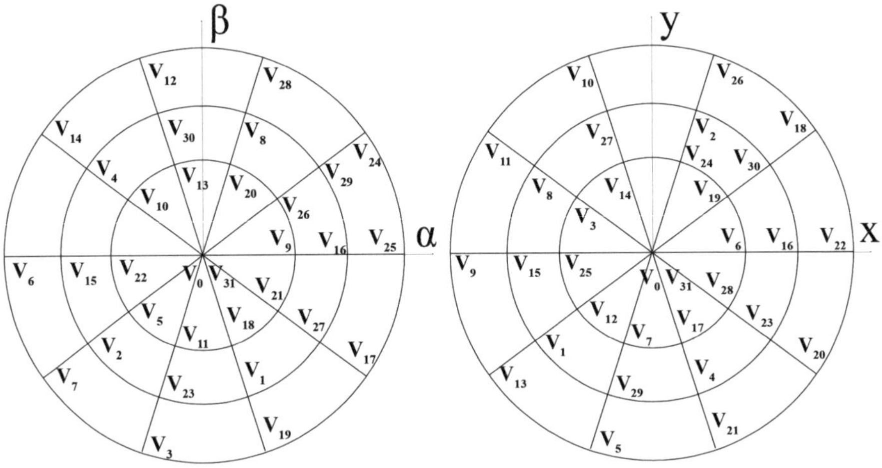

(a) (b)

Figure 3.
Voltage vectors (a) v_β *vs.* v_α *(b)* v_y *vs.* v_x*.*

voltage vector by minimizing a predefined cost function. The number of available voltage vectors for 2L–5Φ inverter is 32 which are 16 if an OPF occurs in any one of the phases. The controller selects an optimum voltage vector by using the discrete mathematical models of the FPIM and inverter. The control objectives such as stator currents (α–β, x–y), stator flux, and torque are predicted by the discrete mathematical models of IM (Eqs. (4)–(12)) and inverter (Eqs. (27)–(29)). The two-steps ahead prediction strategy proposed in [10] is used for predicting the control objective. Hence, the control objectives are predicted for the time instant $k + 2$. The predicted control objectives are then evaluated by a predefined cost function. The cost function is designed based on the torque and flux errors, harmonics components i.e. x–y component, and high current protection constant. The mathematical formulation of the cost function with the two-steps ahead prediction is given below.

$$g = \left|\tau_e^*(k+2) - \tau_e^p(k+2)\right| + \lambda_f\left|\left|\psi_s^*(k+2)\right| - \left|\psi_s^p(k+2)\right|\right| + K + \left|\left|i_{sxy}^*(k+2)\right| - \left|i_{sxy}^p(k+2)\right|\right| \tag{30}$$

where K is a high current protection constant and λ_f is the weighting factor. The harmonic reference current i_{sxy}^* is zero in healthy condition. Note that the cost function (30) must require slight modification in faulty operation. The new i_{sxy}^* are generated by the fault identification sub-section, as shown in **Figure 1**, for the cost function in a faulty condition. In Eq. (30), the reference current i_{sx}^* is set as $-i_{s\alpha}^*$ and the reference current i_{sy}^* is set as zero for minimum copper loss in the faulty situation. The cost function (30) is modified as follows in a faulty condition.

$$g = \left|\tau_e^*(k+2) - \tau_e^p(k+2)\right| + \lambda_f\left|\left|\psi_s^*(k+2)\right| - \left|\psi_s^p(k+2)\right|\right| + K + \left|-i_{s\alpha}^*(k+2) - i_{sx}^p(k+2)\right| + \lambda_h\left|0 - i_{sy}^p(k+2)\right| \tag{31}$$

Where, λ_h is a weighting factor for i_{sy} current. The switching state which yields minimum g is applied to the inverter; this is done by the optimization function, as shown in **Figure 1**. For reliable/safe operation during the fault, the number of motor control variables of the faulty phase is reduced. Hence, the motor intends to run in de-rated capacity by control reconfiguration, and thus the fault-tolerant capability of the drive system is achieved.

3. Mode of operation

An FPIM fed by a 2L–5Φ is controlled by the PTC algorithm, and the motor is operated under both healthy and faulty conditions. The parameters of the motor and controller are given in Appendix (**Table 3**). The performance of the PTC for both healthy and faulty conditions of the motor has been illustrated in the following sub-section.

3.1 Healthy mode

3.1.1 No-load operation

The no-load speed, torque, stator current, stator flux, and fault indices responses under healthy condition are shown in **Figure 4**. The motor is driven at 500 rpm with no load. The stator flux vector is maintained constant at 0.55 wb. It can be seen that the motor behavior is good. The machine takes only 0.4 *sec* to track the reference speed, which is very fast. The estimated developed torque accurately tracks the reference torque and the ripple in the torque response is satisfactory. The stator current

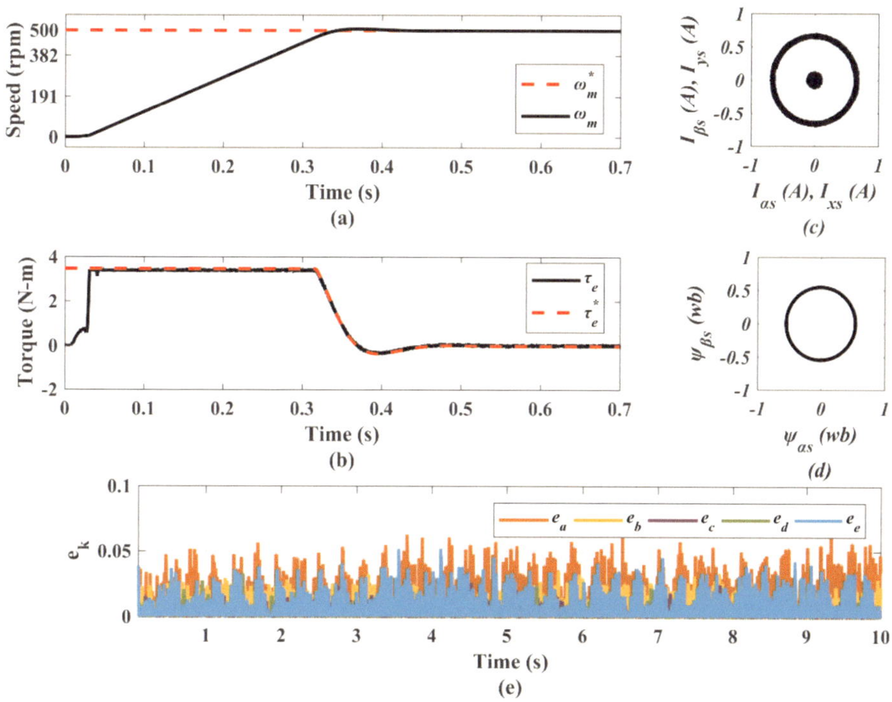

Figure 4.
(a) No-load speed, (b) torque, (c) stator current, (d) stator flux, and (e) fault indices response at no-load under healthy conditions.

and estimated stator flux are presented in the $\alpha-\beta$ plane. The harmonic component of stator current (i.e. $x-y$ component) is represented in the $x-y$ plane which is nearly zero. It is noticed that stator currents and stator flux construct a circular path at the steady-state condition with no load. Fault indices also indicate that each phase is in healthy condition as the fault index of each phase remains below the threshold value of 0.15.

3.1.2 Loaded operation

The loading behavior of the drive system is shown in **Figure 5**. A load of 56% of the nominal load (4.7 N-m) is suddenly applied to the motor at t = 2 sec. There is a tiny speed drop in the motor due to this load disturbance. However, the controller recovers the speed within a short time as shown in **Figure 5(a)**. Hence, the system is robust against load disturbance. The torque ripple is low, as can be seen in **Figure 5(b)**. The stator current and stator flux contain less harmonics and can be seen in **Figures 5(c)** and **5(d)** respectively. As per the healthy condition, fault indices of all phases remain close to zero as shown in **Figure 5(e)**.

3.1.3 Speed reversal operation

The speed transient behavior of the motor drive is shown in **Figure 6**. The motor is driven at 500 rpm, and a reverse speed of −500 rpm is commanded at t = 6 sec with

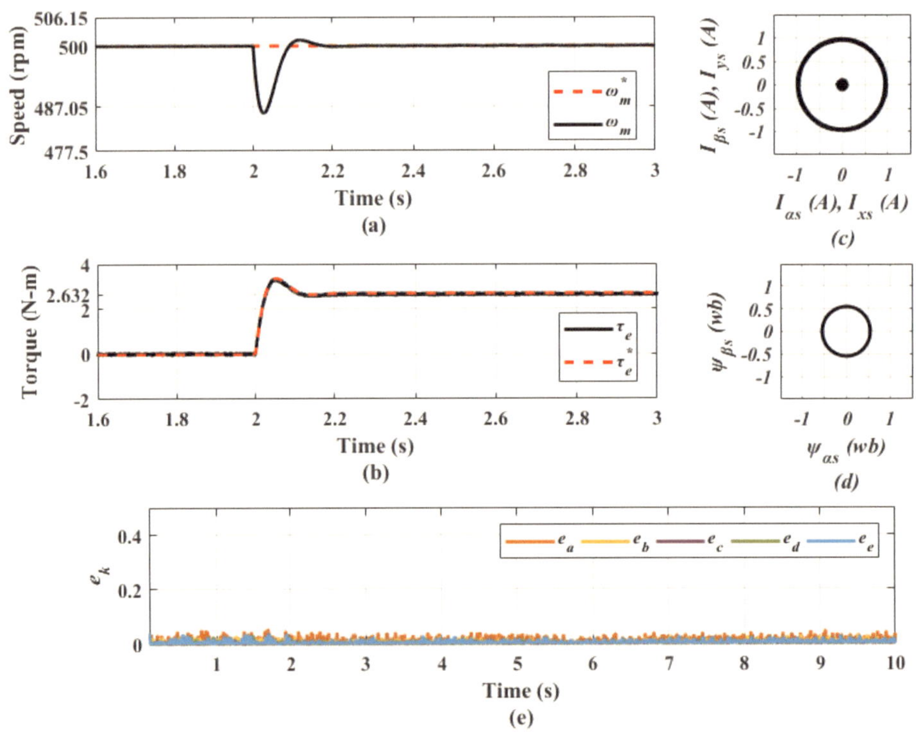

Figure 5.
(a) Speed, (b) torque, (c) stator current, (d) stator flux, and (e) fault indices response at 56% of nominal load under healthy conditions.

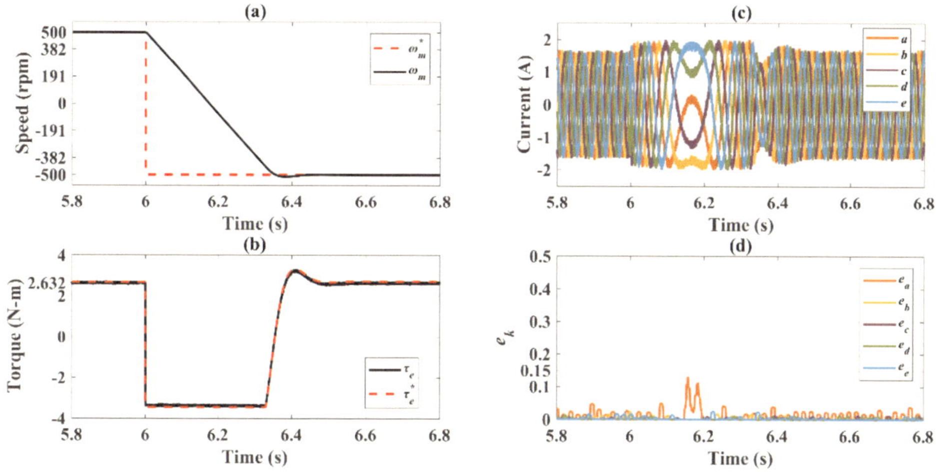

Figure 6.
Speed reversal test (a) speed, (b) torque, (c) current, (d) fault indices response at 56% of nominal load under healthy condition.

56% of nominal torque. The motor follows the command speed quickly without a negligible over/undershoot. Hence, the speed transient behavior of the PTC-based drive is good under healthy condition. From **Figure 6(a)**, it can be seen that the machine takes only 0.5 sec to follow the reverse speed command. **Figure 6(c)** describes that while reversing the direction of rotation, stator currents pass through the condition where for fraction of a second these become DC, and the current in phase '*a*' tends to be zero for a short time interval. So, the controller is bound to detect that there is an open phase fault in phase '*a*' and the fault index of phase '*a*' tries to surpass the threshold value of 0.15 in **Figure 6(d)** but this phenomenon does not stretch out for a long time. So, the fault index again is forced to be maintained at close to zero. Setting a high threshold value of fault indices at speed transient can solve this problem, but it will introduce fault detection delay which may lead the system unstable at faulty situation. Hence, a trade-off is required between the threshold value of fault indices and the fault detection delay.

3.2 Faulty mode

3.2.1 Fault detection

The behavior of the PTC-based FPIM drive is tested in a faulty condition. Only OPF behavior is considered, and the fault is detected using the VSDFD technique. The machine is driven at 500 rpm with 56% of the nominal load, and the VSDFD algorithm is executed from the beginning when the motor was running in healthy condition. When the machine reaches the steady state, phase '*a*' is disconnected from the inverter at t = 4 sec, as shown in **Figure 7**. **Figure 7(a)** justifies the relation which takes place between the x component and α component of stator current (Eq. (15)) when an OPF occurs in phase '*a*'. From **Figure 7(b)**, it can also be seen that the fault indices are almost null in pre-fault or healthy condition, and the fault index 'e_a' increases in the post-fault situation. Notice, all other indices of remaining healthy phases, except 'e_a', in a post-fault situation have similar values (i.e. almost null) as

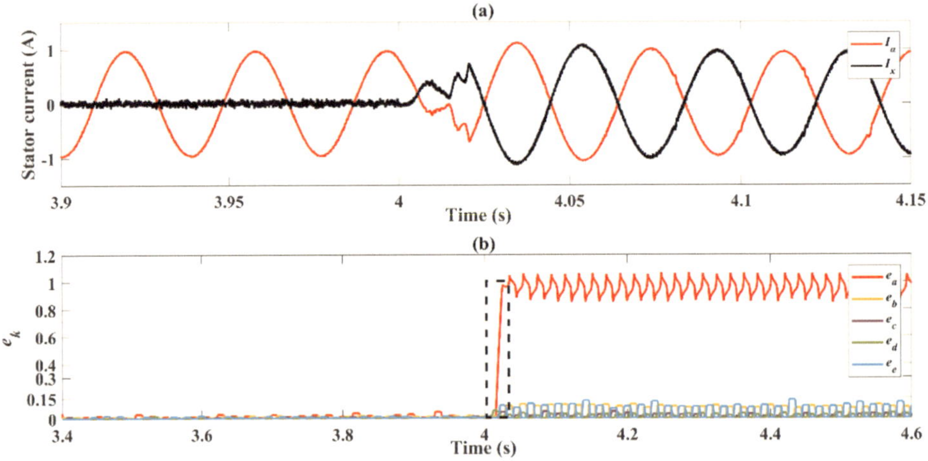

Figure 7.
Open switch fault detection (a) relation between I_α *and* I_x *and (b) fault indices.*

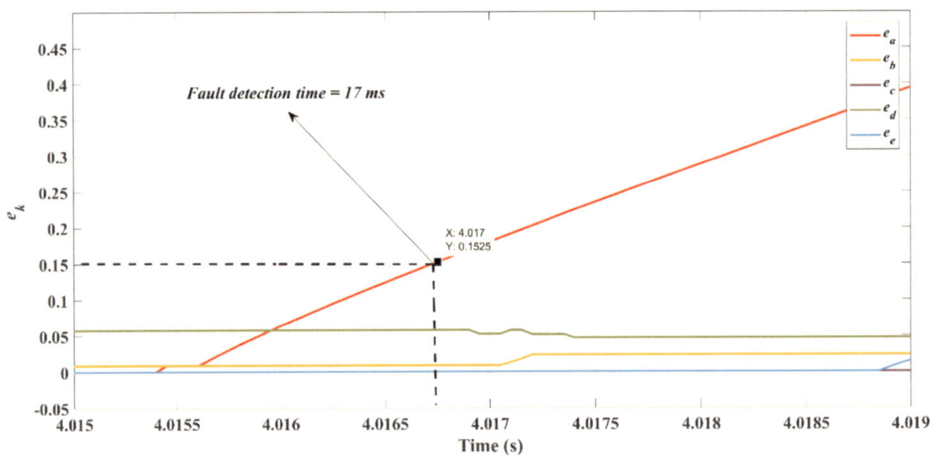

Figure 8.
Zoom-in figure of the fault indices of the marked portion in ***Figure 7(b)****.*

pre-fault situation. A threshold value of indices is set as 0.15 to detect the fault and avoid false alarms as well, and it can be seen in **Figure 8**. Once the value of an index becomes greater than or equal to 0.15, the controller detects a fault in the corresponding phase. In **Figure 8**, it can be seen that the controller takes only 17 ms to detect the fault in phase '*a*', which is 51% of the fundamental current cycle.

3.2.2 Loaded operation

There is a smooth transition from the pre-fault to the post-fault situation of the motor drive, as shown in **Figure 9**. As the fault is injected in phase '*a*' at time t = 4 sec, the phase current 'i_a' is zero in a post-fault condition. The controller tracks the reference speed of 500 rpm quickly in the post-fault situation and also yields similar torque responses in pre-fault and post-fault situations. However, the phase currents

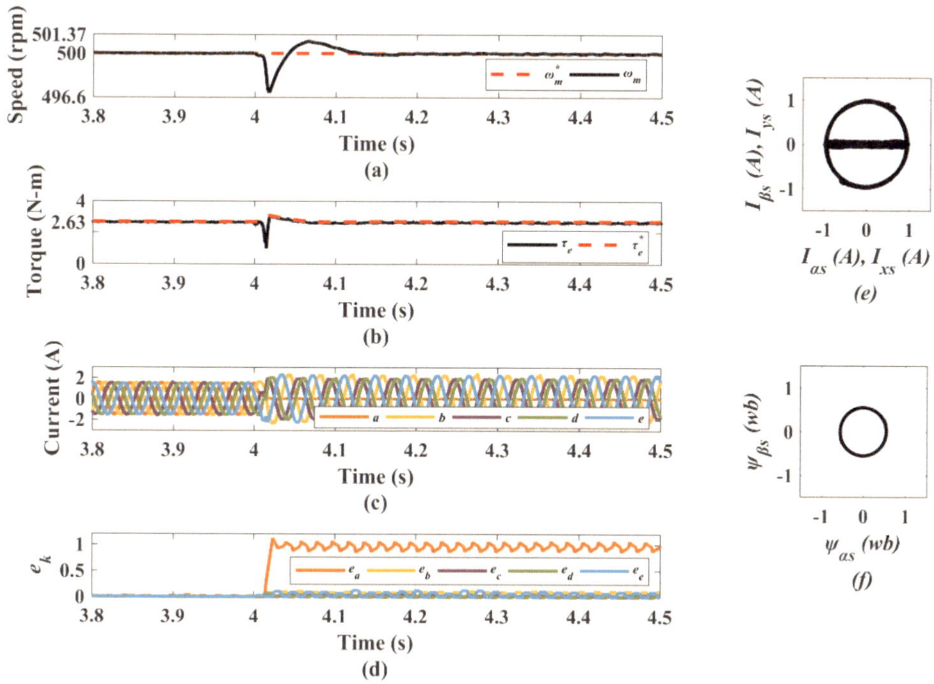

Figure 9.
(a) Speed, (b) torque, (c) current, (d) fault indices response at 56% of nominal load torque in OPF condition.

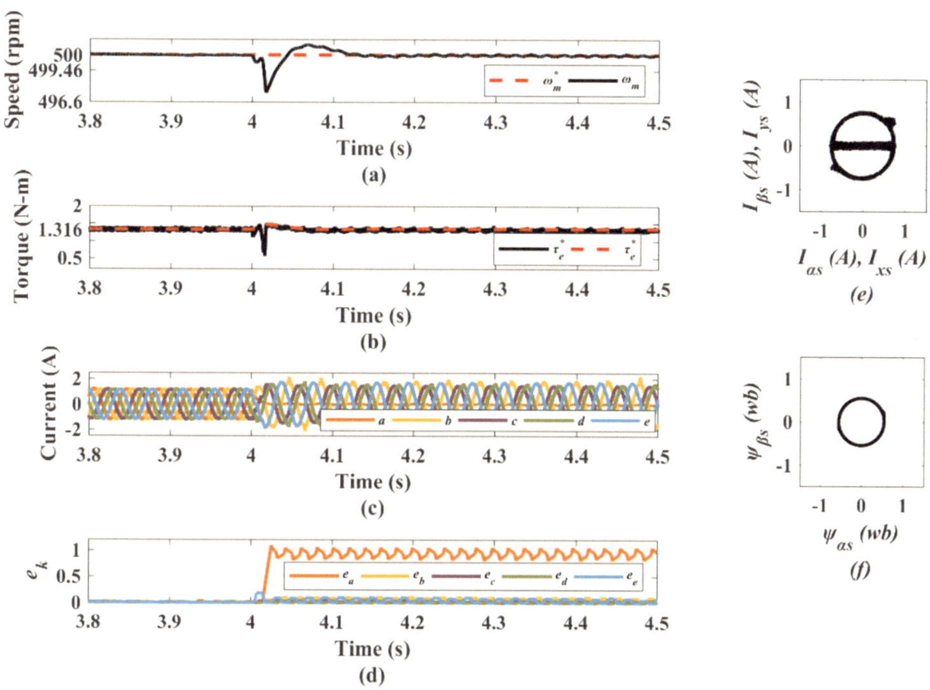

Figure 10.
(a) Speed, (b) torque, (c) current, (d) fault indices response at 28% of nominal load torque in OPF condition.

become unbalanced in the faulty situation. The phase 'b' and phase 'e' currents are the same but higher than the phase 'c' and phase 'd' currents. In order to avoid over-current flow through the stator winding the machine is operated in a de-rated torque condition which is 56% of nominal torque. **Figure 9(e)** shows that the $\alpha-\beta$ current components trace a circle.

However, the $x-y$ current components trace a straight line along the x-axis because the y component is maintained at zero (for minimum copper loss) in the post-fault situation. **Figure 9(f)** shows that the $\alpha-\beta$ flux traces a circle which is expected. Similar speed, torque, current, and flux responses are also achieved with 28% of nominal load torque as shown in **Figure 10**. **Figure 11** shows the speed, torque, current, flux, and fault indices responses of the drive system at 70% of the nominal load torque. It can be seen that the controller cannot track the reference torque properly. It is because there is a stator current rating limit at i_s = 2.3A to protect the stator winding from over-current. However, the controller detects the OPF properly and maintains a smooth transition from pre-fault to post-fault.

3.2.3 Speed reversal operation

Figure 12 demonstrates the rated-speed transient behavior of the machine under OPF situation while the machine carries 56% of the nominal load. **Figure 12(b)** shows that electrical torque is satisfactorily following the reference torque with less ripple. It

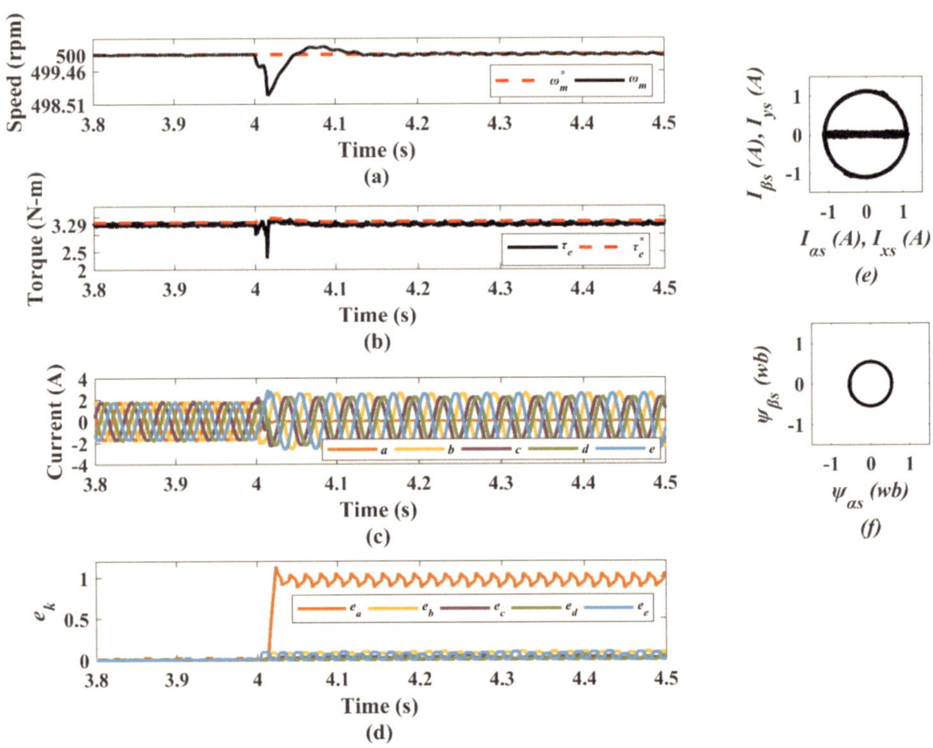

Figure 11.
(a) Speed, (b) torque, (c) current, (d) fault indices response at 70% of nominal load torque in OPF condition.

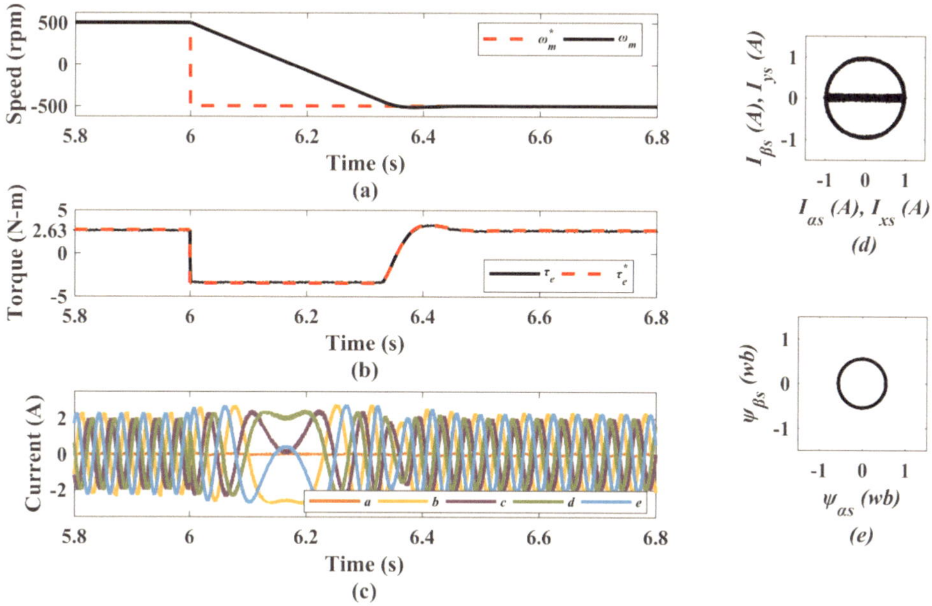

Figure 12.
(a) Speed, (b) torque, (c) stator current and (d) fault indices response at faulty condition under speed reversal with 56% of nominal load.

can also be seen that the stator '*a*' phase current is zero because of OPF. The stator current and the stator flux behavior are also satisfactory as can be seen in **Figure 12(d)** and **(e)** respectively.

4. Conclusions

Predictive torque control (PTC) of a five-phase induction motor (FPIM) with fault-tolerant capability is discussed in this chapter. Both healthy and faulty conditions (i.e. OPF) of the motor drive have been analyzed. The vector space decomposition fault detection (VSDFD) technique is used in PTC to detect the fault. Based on the detected fault, PTC reconfigures its control structure so that it can control the motor in a faulty situation. Basically, the controller sets a new harmonic current reference in the cost function for handling the post-fault situation. The VSDFD algorithm is executed in both healthy and faulty conditions. If there is no fault occurs, the VSDFD algorithm generates an output of zero for each phase, otherwise one. Based on this zero or one output, the controller determines whether it needs to reconfigure itself or not. There are two challenges of a fault detection algorithm that is applicable for multiphase drive: one is fault detection speed and another one is a smooth transition from pre-fault to the post-fault situation. It is shown that the controller can detect the fault within 17 ms which is fast, and this detection time is a fraction of the fundamental current cycle. The proposed PTC can control the motor in both healthy and faulty conditions effectively. Moreover, a smooth transition is maintained from pre-fault to post-fault situations. The controller yields similar speed, torque, stator flux, and stator current responses at any load torque within the rating of the machine in both healthy and faulty conditions. Hence, the proposed PTC for the FPIM drive is fault-tolerant and robust against load disturbance.

Acknowledgements

This research is funded by University Grant Commission (UGC), Bangladesh. I also thank Khulna University of Engineering & Technology (KUET), Khulna, Bangladesh for allowing me to conduct this research.

Nomenclature

θ	phase displacement angle
$\boldsymbol{S_i}$, $i = \{a, b, c, d, e\}$	switching state of each inverter leg
$\boldsymbol{V_{dc}}$	DC link voltage
$\boldsymbol{V_{is}}$, $i = \{a, b, c, d, e\}$	phase voltages
$\boldsymbol{I_s}$, $\boldsymbol{I_r}$	stator and rotor currents
$\boldsymbol{\psi_s}$, $\boldsymbol{\psi_r}$	stator and rotor flux
$\boldsymbol{\tau_e}$	electromagnetic torque
σ	total leakage factor
$\boldsymbol{K_r}$	rotor coupling factor
$\boldsymbol{R_\sigma}$	equivalent resistance ref. to the stator
$\boldsymbol{\tau_\sigma}$	transient stator time constant
$\boldsymbol{L_\sigma}$	leakage inductance
$\boldsymbol{\tau_r}$	rotor time constant
$\boldsymbol{\omega_m}$	rotor angular speed
$\boldsymbol{\omega_e}$	rotor angular frequency
R_i	fault indices

Appendix

Symbols	Variables
V_s	Stator voltage
I_s	Stator current
I_r	Rotor current
ψ_s	Stator flux
ψ_r	Rotor flux
τ_e	Electromagnetic torque
τ_l	Load torque
ω_m	Rotor angular speed
ω_e	Rotor angular frequency
p	Pole pairs

Table 1.
Machine variables.

Factor	Symbols	Expression
Total leakage factor	σ	$1 - \frac{L_m^2}{L_s L_m}$
Rotor coupling factor	τ_e	$\frac{L_m}{L_r}$
Equivalent resistance ref. to the stator	τ_e	$R_s + K_r^2 R_r$
Transient stator time constant	τ_e	$\frac{L_\sigma}{R_\sigma}$
Leakage inductance	τ_e	σL_s
Rotor time constant	τ_e	$\frac{L_r}{R_r}$

Table 2.
Compact factors used in machine modeling.

Parameter	Value	Parameter	Value	Parameter	Value
T_s	50 μs	p	3	V_{dc}	300 V
K_p, K_i	1.09, 38.48	L_m	681.7 mH	J	0.02 kg m^2
λ_f	15	$L_s = L_r$	761.63 mH	ϕ_{nom}	0.55 Wb
λ_h	0.7	R_s	12.85 Ω	T_{nom}	4.7 N-m

Table 3.
Controller and machine parameters used in the simulation.

Author details

Md. Habibullah*, Tusar Debanath and Md. Shahadath Hossain Sabbir
Department of Electrical and Electronic Engineering, Khulna University of Engineering and Technology, Khulna, Bangladesh

*Address all correspondence to: habibullah@eee.kuet.ac.bd

References

[1] Riveros JA, Barrero F, Levi E, Duran MJ, Toral S, Jones M. Variable-speed five-phase induction motor drive based on predictive torque control. IEEE Transactions on Industrial Electronics. 2013;**60**(8):2957-2968. DOI: 10.1109/TIE.2012.2198034

[2] Bermudez M, Gonzalez-Prieto I, Barrero F, Guzman H, Duran MJ, Kestelyn X. Open-phase fault-tolerant direct torque control technique for five-phase induction motor drives. IEEE Transactions on Industrial Electronics. 2017;**64**(2):902-911. DOI: 10.1109/TIE.2016.2610941

[3] Guzman H, et al. Comparative study of predictive and resonant controllers in fault-tolerant five-phase induction motor drives. IEEE Transactions on Industrial Electronics. 2016;**63**(1):606-617. DOI: 10.1109/TIE.2015.2418732

[4] Guzman H, Duran MJ, Barrero F, Bogado B, Toral S. Speed control of five-phase induction motors with integrated open-phase fault operation using model-based predictive current control techniques. IEEE Transactions on Industrial Electronics. 2014;**61**(9):4474-4484. DOI: 10.1109/TIE.2013.2289882

[5] González-Prieto I, Duran MJ, Rios-Garcia N, Barrero F, Martín C. Open-switch fault detection in five-phase induction motor drives using model predictive control. IEEE Transactions on Industrial Electronics. 2018;**65**(4):3045-3055. DOI: 10.1109/TIE.2017.2748052

[6] Barrero F, Bermudez M, Duran MJ, Salas P, Gonzalez-Prieto I. Assessment of a Universal Reconfiguration-less Control Approach in Open-Phase Fault Operation for Multiphase Drives. Energies 2019;**12**:4698. DOI: 10.3390/en12244698

[7] Zhao Y, Lipo TA. Modeling and control of a multi-phase induction machine with structural unbalance. IEEE Transactions on Energy Conversion. 1996;**11**(3):578-584. DOI: 10.1109/60.537028

[8] Duran MJ, Gonzalez-Prieto I, Rios-Garcia N, Barrero F. A simple, fast, and robust open-phase fault detection technique for six-phase induction motor drives. IEEE Transactions on Power Electronics. 2018;**33**(1):547-557. DOI: 10.1109/TPEL.2017.2670924

[9] Chikondra B, Muduli UR, Behera RK. An improved open-phase fault-tolerant DTC technique for five-phase induction motor drive based on virtual vectors assessment. IEEE Transactions on Industrial Electronics. 2021;**68**(6):4598-4609. DOI: 10.1109/TIE.2020.2992018

[10] Cortes P, Rodriguez J, Silva C, Flores A. Delay compensation in model predictive current control of a three-phase inverter. IEEE Transactions on Industrial Electronics. 2012;**59**(2):1323-1325. DOI: 10.1109/TIE.2011.2157284

Chapter 7

Mathematical Modeling of a Three-Phase Induction Motor

Abdelkarim Belbali, Salim Makhloufi, Abdellah Kadri, Laidi Abdallah and Zemitte Seddik

Abstract

The induction machine (IM) due to its simplicity of design and maintenance has been favored by manufacturers since its invention by N. Tesla, when he discovered the rotating magnetic fields generated by a system of polyphase currents. However, this simplicity reaches great physical complexity, related to the electromagnetic interactions between the stator and the rotor, which is why it has long been used in constant-speed drives. The induction machine is currently the most widely used electric machine in the industry. Its main advantages lie in the absence of rotor winding, simple structure, robust, and easy to build. A mathematical model is used to represent or reproduce a given real system. The interest of a model is the analysis and prediction of the static and dynamic behavior of the physical system. This chapter's goals are to provide an overview of a three-phase induction machine's mathematical model, and its transformation into the two-phase (α, β) Concordia system.

Keywords: induction machine (IM), Concordia system, mathematical model, Clarke transformation, park transformation

1. Introduction

Induction machine (IM) is widely used in industrial applications. Indeed, due to its design, its cost is low compared to that of other machines. It is also very robust under different conditions of use. However, the relative simplicity of the machine's design hides a great functional complexity.

IM depending on whether it is wound rotor or squirrel cage contains a stator and a rotor, made up of silicon steel sheets stack, and containing notches in which the windings are placed, the latter being arranged in such a way that, when supplied by a three-phase electric power produces a rotating field at the frequency of the power supply. This rotating field results in the generation of eddy currents (also called Foucault's currents) in the rotor bars where a large force results from the interaction of the stator and rotor magnetic fields causing the torque to be generated. However, the squirrel cage structure is often taken during modeling as electrically equivalent to that of a wound rotor whose windings are short-circuited.

The objective of this chapter is to present mathematically the modeling of the induction machine in the form of different state models according to the chosen reference frame. These models are defined in a two-phase frame of reference, either rotating (dq), or fixed to the stator $(\alpha\beta)$, the latter is determined from the conventional three-phase reference frame of the induction machine using suitable mathematical transformations.

2. Operating principle

To operate the machine in motor mode, the rotor must be rotated in the direction of the rotating magnetic field, at a speed lower than the synchronous speed (the speed of the rotating field), that is expressed by the following equation [1]:

$$\Omega_s = \frac{60f}{p} \tag{1}$$

with

Ω_s: synchronism speed;

f: Electric Network Frequency (ENF);

p: number of pole pairs.

The speed at which this machine begins to operate (motor mode operation) when it is linked to the electrical network is just a little bit slower than the speed of the stator magnetic field [2]. If the rotational speed of the rotor becomes the same (synchronous) as that of the magnetic field, no induction appears in the rotor; therefore, no interaction happens with the stator (motor stopped) [3]. Finally, if the rotation speed of the rotor is slightly higher than that of the stator magnetic field (generator mode operation), an electromagnetic force similar to that obtained with a synchronous generator will be developed [4]. The difference between the rotation speed of the rotor and that of the magnetic field is called the slip [5], and practically its value does not exceed few percent.

However, from a certain rotational speed, a noticeable decrease in the motor's stator flux occurs, which requires more current for a similar torque. After reaching a maximum torque value, a reduction in torque and consequently electrical power is observed. **Figure 1** illustrates the induction motor components [6].

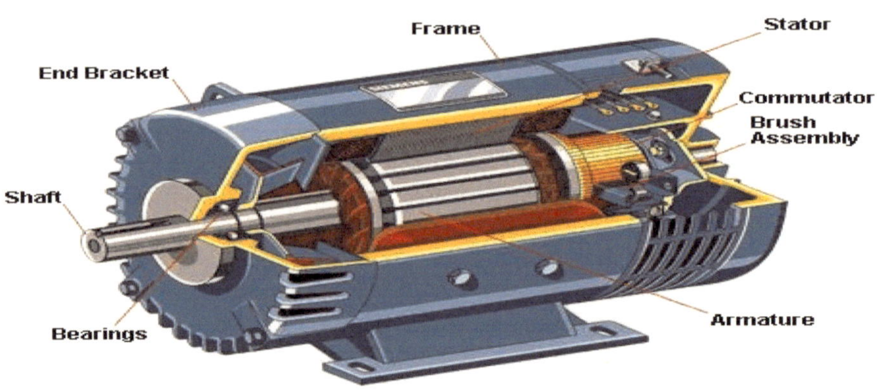

Figure 1.
Induction motor components [6].

3. Induction machine modeling

3.1 Mathematical model of the induction motor

Modeling of any physical system is necessary in the research filed because it allows researchers to predict how the system can be improved against various phenomena, and thus, learn more about the mechanisms that control it. The induction machine can be modeled by different methods, depending on the desired purposes. The following models are developed in this chapter:

- Models in *abc* frame, resulting from differential equations controlling the operation of the machine. They are used mainly for the steady-state study [7, 8].
- The models resulting from Concordia's transformation are commonly used for the dynamic-state study and for the direct torque control (DTC) [9].

3.2 Simplifying assumptions

An induction machine, with its windings distribution and geometry, is so complex that it cannot be analyzed, taking into account its exact configuration. Then, it is necessary to adopt simplifying assumptions [10, 11]:

- The constant air gap;
- The neglected notching effect;
- Sinusoidal spatial distribution of magneto-motor air forces;
- Unsaturated magnetic circuit with constant permeability;
- Negligible ferromagnetic losses;
- The skin effect and warming effect on the characteristics are not taken into account.

Among the important consequences of these assumptions are.

- The association of flux;
- The self-inductances constancy;
- The invariance of stator resistances and rotor resistances;
- The sinusoidal variation law of the mutual inductances between the stator and rotor windings in terms of the electric angle of their magnetic axes;

The induction machine is represented schematically by **Figure 2**. It has six windings:

- The machine stator consists of three fixed windings shifted by 120° in space and crossed by three variable currents.

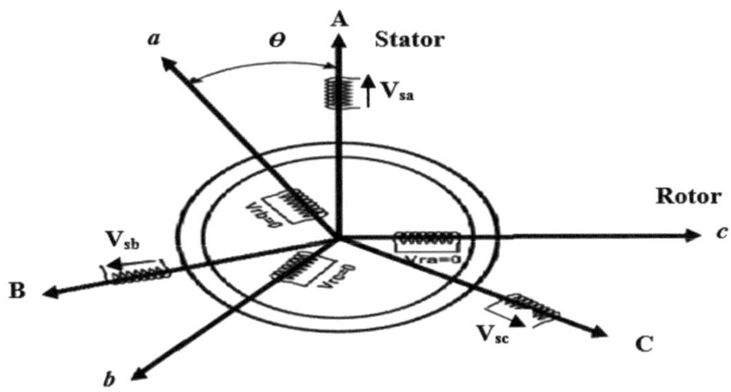

Figure 2.
Representation of the induction machine.

- The rotor can be modeled by three identical windings shifted in space by 120°. These windings are short circuits and the voltage across them is zero.

3.3 Induction machine modeling

Aforesaid, to ensure motor operation, the IM's rotation speed must be lower than the synchronization speed (positive slip). Unlike the synchronous machine, the IM does not have a separate inductor; therefore, it requires a reactive power input for its magnetization. When it connected directly to the grid, the latter provides the required reactive power. On the other hand, in autonomous operation, it is necessary to bring this energy either by a battery of capacitors or by a controlled static converter (an inverter).

A mathematical model is necessary for the analysis of the IM's operation in both motor and generator modes. The analytical modeling will be presented in the section below.

3.3.1 Electrical equations of the induction machine in the three-phase reference

The induction machine is of three-phase nature. Taking into account the assumptions mentioned above and using the diagram shown in **Figure 3**, the induction machine's basic equations are [12, 13]:

$$[v_{sabc}]^T = [R_s][i_{sabc}]^T + \frac{d}{dt}[\varphi_{sabc}]^T \tag{2}$$

$$[v_{rabc}]^T = 0 = [R_r][i_{rabc}]^T + \frac{d}{dt}[\varphi_{rabc}]^T \tag{3}$$

With.

v_{sabc}: The voltages applied to the three-stator phases.

i_{sabc}: The currents that cross the three-stator phases.

φ_{sabc}: The total flux through these windings.

R_s: The stator resistance.

R_r: The rotor resistance.

Each flux comprises an interaction with the currents of all the phases including its own.

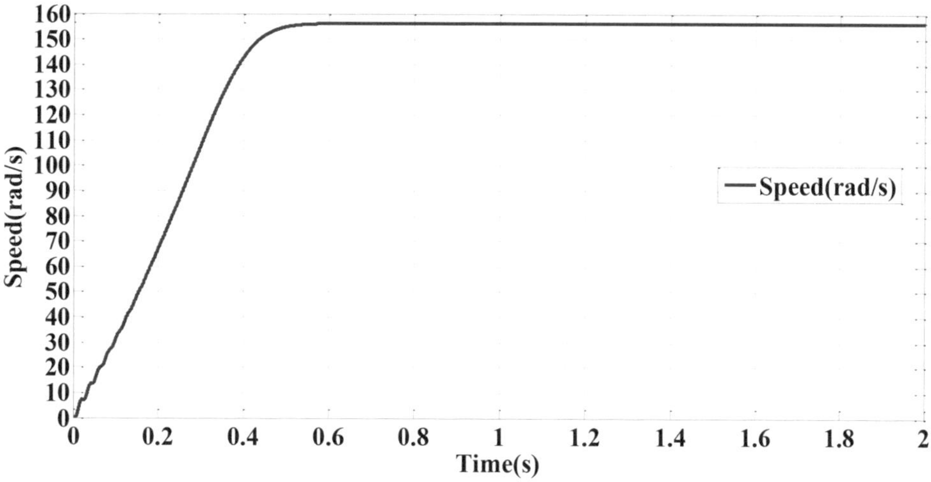

Figure 3.
Rotor speed simulation result.

$$\begin{bmatrix} \varphi_{sa} \\ \varphi_{sb} \\ \varphi_{sc} \\ \varphi_{ra} \\ \varphi_{rb} \\ \varphi_{rc} \end{bmatrix} = \begin{bmatrix} l_s & m_s & m_s & m_1 & m_3 & m_2 \\ m_s & l_s & m_s & m_2 & m_1 & m_3 \\ m\_s & m_s & l_s & m_3 & m_2 & m_1 \\ m_1 & m_2 & m_3 & l_r & m_r & m_r \\ m_3 & m_1 & m_2 & m_r & l_r & m_r \\ m_2 & m_3 & m_1 & m_r & m_r & l_r \end{bmatrix} \begin{bmatrix} i_{sa} \\ i_{sb} \\ i_{sc} \\ i_{ra} \\ i_{rb} \\ i_{rc} \end{bmatrix} \tag{4}$$

where.
l_s is the self-inductance of a stator phase.
l_r is the self-inductance of a rotor phase.
m_s is the mutual inductance between two stator phases.
m_r is the mutual inductance between two rotor phases.
m_{sr} is the maximum mutual inductance between a stator phase and a rotor phase.

$$m_1 = m_{sr}\cos(\theta) \tag{5}$$

$$m_2 = m_{sr}\cos\left(\theta - \frac{2\pi}{3}\right) \tag{6}$$

$$m_3 = m_{sr}\cos\left(\theta + \frac{2\pi}{3}\right) \tag{7}$$

3.3.2 Three-phase/two-phase transformation (Concordia and Clarke transformation)

The aim of using this transformation is to switch from a three-phase *abc* system to the stationary two-phase $\alpha\beta$ system [14–16]. There are mainly two transformations:

Clarke and Concordia transformations. The magnitude of the converted quantities is saved by Clarke transformation, but neither the power nor the torque is (we must multiply by a coefficient of 3/2) [17]. While Concordia transformations keeps the power but not the magnitude (**Tables 1** and **2**) [18].

3.3.3 Three-phase/two-phase transformation (park transformation)

Park transformation is a transformation of the fixed three-phase reference frame relative to the stator in a two-phase reference frame [19]. It allows to move from the *abc* reference to the (d, q) reference, where d refers to the direct axis and q to the quadrature axis. The (α, β) reference frame is always fixed according to the abc frame [20], where the (d, q) reference frame is mobile [21].

This transformation reduces the complexity of the system. The reference frame position can be fixed according to the three referential [22, 23]:

- Reference system linked to the rotating field.
- Referential linked to the stator.
- Reference system linked to the rotor.

Concordia transformation	**Clarke transformation**
$\begin{bmatrix} x_a \\ x_b \\ x_c \end{bmatrix} \overset{T_{23}}{\rightarrow} \begin{bmatrix} x_\alpha \\ x_\beta \end{bmatrix}$ i.e. $[x_{\alpha\beta}]^T = T_{23}[x_{abc}]^T$	$\begin{bmatrix} x_a \\ x_b \\ x_c \end{bmatrix} \overset{C_{23}}{\rightarrow} \begin{bmatrix} x_\alpha \\ x_\beta \end{bmatrix}$ i.e. $[x_{\alpha\beta}]^T = C_{23}[x_{abc}]^T$
with: $T_{23} = \sqrt{\frac{2}{3}} \begin{bmatrix} 1 & -\frac{1}{2} & -\frac{1}{2} \\ 0 & \frac{\sqrt{3}}{2} & -\frac{\sqrt{3}}{2} \end{bmatrix}$	with: $C_{23} = \frac{2}{3} \begin{bmatrix} 1 & -\frac{1}{2} & -\frac{1}{2} \\ 0 & \frac{\sqrt{3}}{2} & -\frac{\sqrt{3}}{2} \end{bmatrix}$

Table 1.
Concordia and Clarke transformations.

Inverse Concordia transformation	**Inverse Clarke transformation**
$\begin{bmatrix} x_\alpha \\ x_\beta \end{bmatrix} \overset{T_{32}}{\rightarrow} \begin{bmatrix} x_a \\ x_b \\ x_c \end{bmatrix}$ i.e. $[x_{abc}]^T = T_{32}[x_{\alpha\beta}]^T$	$\begin{bmatrix} x_\alpha \\ x_\beta \end{bmatrix} \overset{C_{32}}{\rightarrow} \begin{bmatrix} x_a \\ x_b \\ x_c \end{bmatrix}$ i.e. $[x_{\alpha\beta}]^T = C_{32}[x_{abc}]^T$
with: $T_{32} = \sqrt{\frac{2}{3}} \begin{bmatrix} 1 & 0 \\ -\frac{1}{2} & \frac{\sqrt{3}}{2} \\ -\frac{1}{2} & -\frac{\sqrt{3}}{2} \end{bmatrix}$	with: $C_{32} = \begin{bmatrix} 1 & 0 \\ -\frac{1}{2} & \frac{\sqrt{3}}{2} \\ -\frac{1}{2} & -\frac{\sqrt{3}}{2} \end{bmatrix}$

Table 2.
Transformation of a two-phase $\alpha\beta$ system to a three-phase abc (system).

The transformation matrix of Park and its inverse are given by

$$\begin{cases} P(\theta)=k\begin{pmatrix} \cos(\theta) & \cos\left(\theta-\frac{2\pi}{3}\right) & \cos\left(\theta+\frac{2\pi}{3}\right) \\ -\sin(\theta) & -\sin\left(\theta-\frac{2\pi}{3}\right) & -\sin\left(\theta+\frac{2\pi}{3}\right) \\ \frac{1}{\sqrt{2}} & \frac{1}{\sqrt{2}} & \frac{1}{\sqrt{2}} \end{pmatrix} \\ P(\theta)^{-1}=P(\theta)^{T}=k\begin{pmatrix} \cos(\theta) & -\sin(\theta) & \frac{1}{\sqrt{2}} \\ \cos\left(\theta-\frac{2\pi}{3}\right) & -\sin\left(\theta-\frac{2\pi}{3}\right) & \frac{1}{\sqrt{2}} \\ \cos\left(\theta+\frac{2\pi}{3}\right) & -\sin\left(\theta+\frac{2\pi}{3}\right) & \frac{1}{\sqrt{2}} \end{pmatrix} \end{cases} \tag{8}$$

where k is a constant that can take the value $2/3$ for the transformation with no power conservation or $\sqrt{2/3}$ for the transformation with power conservation [24].

3.4 Model of the induction machine in the park referential

The Park transformation consists in applying to the currents, voltages, and flux a change of variables involving the angle between the windings axis and the axis of the Park (d, q) frame [25].

Eqs. (2), (3), and (4) give rise to the following system [26]:

$$\begin{cases} v_{sd}=R_s i_{sd}+\dfrac{d\phi_{sd}}{dt}-\omega_s\varphi_{sq} \\ v_{sq}=R_s i_{sq}+\dfrac{d\phi_{sq}}{dt}+\omega_s\varphi_{sd} \end{cases} \tag{9}$$

$$\begin{cases} v_{rd}=0=R_r i_{rd}+\dfrac{d\phi_{rd}}{dt}-\omega_r\varphi_{rq} \\ v_{rq}=0=R_r i_{rq}+\dfrac{d\phi_{rq}}{dt}+\omega_r\varphi_{rd} \end{cases} \tag{10}$$

with

$$\begin{bmatrix} \varphi_{sdq} \\ \varphi_{rdq} \end{bmatrix}=\begin{bmatrix} L_s & 0 & L_m & 0 \\ 0 & L_s & 0 & L_m \\ L_m & 0 & L_r & 0 \\ 0 & L_m & 0 & L_r \end{bmatrix}\begin{bmatrix} i_{sdq} \\ i_{rdq} \end{bmatrix} \tag{11}$$

Indeed, the submatrices are now diagonal and no longer depend on θ (the electrical angle between the stator and the rotor).

The matrix system can also be written as [27]:

$$\begin{cases} \varphi_{sd} = L_s i_{sd} + L_m i_{rd} \\ \varphi_{sq} = L_s i_{sq} + L_m i_{rq} \end{cases} \tag{12}$$

$$\begin{cases} \varphi_{rd} = L_m i_{sd} + L_r i_{rd} \\ \varphi_{rq} = L_m i_{sq} + L_r i_{rq} \end{cases} \tag{13}$$

(where)

$$L_s = l_s - m_s, L_r = l_r - m_r, L_m = \frac{2}{3} m_{sr}$$

We have expressed the machine's equations, but there also remains the electromagnetic torque. The latter can be derived from the co-energy expression, or obtained using power balance.

The instantaneous power supplied to the stator and rotor windings is written as [28]

$$P_e = [V_s]^T [I_s] + [V_r]^T [I_r] \tag{14}$$

By applying Park transformation, it is expressed in terms of the axes quantity dq

$$P_e = \begin{bmatrix} v_{sd} & v_{sq} \end{bmatrix} \begin{bmatrix} i_{sd} \\ i_{sq} \end{bmatrix} + \begin{bmatrix} v_{rd} & v_{rq} \end{bmatrix} \begin{bmatrix} i_{rd} \\ i_{rq} \end{bmatrix} = \underbrace{\frac{2}{3}\left[i_{sd}\frac{d\varphi_{sd}}{dt} + i_{sq}\frac{d\varphi_{sq}}{dt} + i_{rd}\frac{d\varphi_{rd}}{dt} + i_{rq}\frac{d\varphi_{rq}}{dt} \right]}_{\textit{first term}}$$

$$\underbrace{+\frac{2}{3}\left[\left(\varphi_{sd} i_{sq} - \varphi_{sq} i_{sd} \right) \omega_s + \left(\varphi_{rq} i_{rd} - \varphi_{rd} i_{rq} \right) \omega_r \right]}_{\textit{second term}} \tag{15}$$

$$\underbrace{+\frac{2}{3}\left[R_s \left(i_{sd}^2 + i_{sq}^2 \right) + R_r \left(i_{rd}^2 + i_{rq}^2 \right) \right]}_{\textit{third term}}$$

- The first term represents the magnetic energy stored in iron.
- The second term represents the electromechanical power P_{em} of the machine.
- The third term represents joule losses.

Taking into account the flux Eqs. (12) and (13), several equal expressions result

$$\begin{cases} \Gamma_{em} = p\left(\varphi_{sd} i_{sq} - \varphi_{sq} i_{sd} \right) \\ \Gamma_{em} = p\left(\varphi_{rq} i_{rd} - \varphi_{rd}\, i_{rq} \right) \\ \Gamma_{em} = p L_m \left(i_{sq} i_{rd} - i_{sd} i_{rq} \right) \\ \Gamma_{em} = p \frac{L_m}{L_r} \left(\varphi_{rd} i_{sq} - \varphi_{rq} i_{sd} \right) \end{cases} \tag{16}$$

where p is the number of poles pairs. The power P_{em} is also equal to $\Gamma_{em}\omega_r/p$, and the movement equation is [29]

$$\Gamma_{em} - \Gamma_r = f\Omega_m + J\frac{d\Omega_m}{dt} \tag{17}$$

3.5 Selecting the *dq* (frame)

The machine's equations and electrical quantities have been expressed thus far in a reference dq, which makes an electrical angle θ_s and θ_r with the stator and with the rotor respectively, but which is not defined elsewhere, that is, it is free [30, 31].

According to the application purpose, there are three main choices for the axis (d, q) frame orientation: a frame linked to the stator, rotor, or linked to the rotating field [32, 33]. In each of these referential, the equations of the machine become simpler than in any other referential [34]. Generally, the operating conditions will typically determine the most convenient reference for analysis and/or simulation purposes.

3.5.1 Reference linked to the stator

Regarding the stator, this referential is immobile. It is carried out to investigate machine braking and starting (i.e., this reference frame is better adapted to work with instantaneous quantities) [35]. In addition, this choice is used for direct torque control design [36].

It is characterized by.

$\omega = \omega_s = 0$ and therefore, $\omega_r = -\omega_m$ (where ω is the arbitrary frame rate).

The system of equations in this reference frame is [37–39]

$$\begin{cases} v_{sd} = R_s i_{sd} + \dfrac{d\varphi_{sd}}{dt} \\ v_{sq} = R_s i_{sq} + \dfrac{d\varphi_{sq}}{dt} \\ v_{rd} = 0 = R_r i_{rd} + \dfrac{d\varphi_{rd}}{dt} - \omega_r \varphi_{rq} \\ v_{rq} = 0 = R_r i_{rq} + \dfrac{d\varphi_{rq}}{dt} + \omega_r \varphi_{rd} \end{cases} \tag{18}$$

3.5.2 Reference linked to the rotor

In the case where the (d, q) reference frame is synchronized with the rotor $\omega = \omega_s = \omega_m$ and $\omega_r = 0$. This reference frame is used for the simulation of the dynamic state of machines where the speed is assumed constant [40]. In this case, the system of equations is [41]

$$\begin{cases} v_{sd} = R_s i_{sd} + \dfrac{d\varphi_{sd}}{dt} - \omega_s \varphi_{sq} \\ v_{sq} = R_s i_{sq} + \dfrac{d\varphi_{sq}}{dt} + \omega_s \varphi_{sd} \\ v_{rd} = 0 = R_r i_{rd} + \dfrac{d\varphi_{rd}}{dt} \\ v_{rq} = 0 = R_r i_{rq} + \dfrac{d\varphi_{rq}}{dt} \end{cases} \tag{19}$$

3.5.3 Reference linked to the rotating magnetic field

This choice allows obtaining a sliding pulsation and properly adapts vector control through rotor flux orientation [42]. The reference frame linked to the synchronism (or rotating field) is fixed relative to the rotating field. It is used for the machine vector control and it is characterized by $\omega = \omega_s$, which implies that the adjustment variables are continuous [43]. The advantage of using this reference frame is to have constant quantities in steady state; then, it is easier to carry out the regulation [44]. Then, we can write [45]

$$\begin{cases} v_{sd} = R_s i_{sd} + \dfrac{d\varphi_{sd}}{dt} - \omega_s \varphi_{sq} \\ v_{sq} = R_s i_{sq} + \dfrac{d\varphi_{sq}}{dt} + \omega_s \varphi_{sd} \\ v_{rd} = 0 = R_r i_{rd} + \dfrac{d\varphi_{rd}}{dt} - \omega_r \varphi_{rq} \\ v_{rq} = 0 = R_r i_{rq} + \dfrac{d\varphi_{rq}}{dt} + \omega_r \varphi_{rd} \end{cases} \tag{20}$$

These equations can be rewritten to have a different state vector (state variables system), that is, instead of having the flux, we can write it in currents; we just need to make substitutions of Eqs. (12) and (13) in Eq. (20).

3.6 Model of the induction machine in the (α, β) (frame)

The dynamic model of an induction motor can be developed from its basic electrical and mechanical equations [46]. In the stationary reference frame, the voltages are expressed as follows [47]:

$$\begin{cases} v_{s\alpha} = R_s i_{s\alpha} + s\varphi_{s\alpha} \\ v_{s\beta} = R_s i_{s\beta} + s\varphi_{s\beta} \\ v_{r\alpha} = 0 = R_r i_{r\alpha} + s\varphi_{r\alpha} + \omega_r \varphi_{r\beta} \\ v_{r\beta} = 0 = R_r i_{r\beta} + s\varphi_{r\beta} - \omega_r \varphi_{r\alpha} \end{cases} \tag{21}$$

where s indicates the differential operator (d/dt).
The stator and rotor fluxes equations are [48]

$$\begin{cases} \varphi_{s\alpha} = L_s i_{s\alpha} + L_m i_{r\alpha} \\ \varphi_{s\beta} = L_s i_{s\beta} + L_m i_{r\beta} \\ \varphi_{r\alpha} = L_r i_{r\alpha} + L_m i_{s\alpha} \\ \varphi_{r\beta} = L_r i_{r\beta} + L_m i_{s\beta} \end{cases} \tag{22}$$

In these equations, R_s, R_r, L_s, and L_r are, respectively, the resistors and the inductances of the stator windings and the rotor windings, L_m is the mutual inductance and $\omega_r = p.\Omega_r$ is the rotor speed (with p is the pairs poles number). Additionally, ω_s is the synchronous pulsation.

$v_{s\alpha}, v_{s\beta}, v_{r\alpha}, v_{r\beta}, i_{s\alpha}, i_{s\beta}, i_{r\alpha}, i_{r\beta}, ø_{s\alpha}, ø_{s\beta}, ø_{r\alpha}$, and $ø_{r\beta}$ are the direct and quadratic components, respectively, of the voltages and currents as well as the fluxes of both the stator and the rotor.

The mechanical equation is [29]

$$\Gamma_{em} - \Gamma_r = f\Omega_m + J\frac{d\Omega_m}{dt} \tag{23}$$

where Γ_{em} is the electromagnetic torque [N.m] and Γ_r is the resistive torque imposed by the machine shaft [N. m].

The electromagnetic torque is [49]

$$\Gamma_{em} = \frac{3}{2}p\left(\varphi_{s\alpha}i_{s\beta} - \varphi_{s\beta}i_{s\alpha}\right) \tag{24}$$

For the complete model of the induction machine, the flux expressions are replaced in the voltage equations. We obtain a mechanical equation and four electrical equations in terms of the stator currents, rotor fluxes components, and the electric speed of induction machine as well [45]:

$$\begin{cases} \dfrac{di_{s\alpha}}{dt} = -\dfrac{1}{\sigma L_s}\left(R_s + \dfrac{1}{T_r}\dfrac{L_m^2}{L_r}\right)i_{s\alpha} + \dfrac{1}{\sigma L_s}\left(\dfrac{L_m}{L_r}\dfrac{1}{T_r}\right)\varphi_{r\alpha} + \dfrac{1}{\sigma L_s}\left(\dfrac{L_m}{L_r}\right)\omega_r\varphi_{r\beta} \\ \dfrac{di_{s\beta}}{dt} = -\dfrac{1}{\sigma L_s}\left(R_s + \dfrac{1}{T_r}\dfrac{L_m^2}{L_r}\right)i_{s\beta} - \dfrac{1}{\sigma L_s}\left(\dfrac{L_m}{L_r}\right)\omega_r\varphi_{r\alpha} + \dfrac{1}{\sigma L_s}\left(\dfrac{L_m}{L_r}\dfrac{1}{T_r}\right)\varphi_{r\beta} \\ \dfrac{d\varphi_{r\alpha}}{dt} = \dfrac{L_m}{T_r}i_{s\alpha} - \dfrac{1}{T_r}\varphi_{r\alpha} - \omega_r\varphi_{r\beta} \\ \dfrac{d\varphi_{r\beta}}{dt} = \dfrac{L_m}{T_r}i_{s\alpha} + \omega_r\varphi_{r\alpha} - \dfrac{1}{T_r}\varphi_{r\beta} \end{cases} \tag{25}$$

Such as $\omega_m = p\,\Omega_m; \omega_r = [\omega_s - \omega_m]; \sigma = 1 - \frac{L_m^2}{L_s L_r}; T_r = \frac{L_r}{R_r}; T_s = \frac{L_s}{R_s}$.

Modeling the machine in this way that reduces the number of quantities that we need to know in order to simulate machine operation. In fact, only the instantaneous values of the stator voltages and the resistive torque must be determined in order to impose them on the machine. Therefore, we do not need to know the stator pulsation value, or the slip as in the case of the model whose equations are written in the reference frame rotating in synchronism [50].

3.7 Voltage powered machine state space representation

The state space representation of the induction machine depends on the selected frame and the selection of state variables for the electrical equations. We write the equations in the (α, β) frame because it is the most general and complete solution [51]. The objectives for either the control or the observation determine the state variables to be used [52].

For a three-phase IM powered by voltage, the stator voltages $(v_{s\alpha}, v_{s\beta})$ are considered as control variables, the load torque Γ_r as a disturbance [53]. In our case, we choose the state vector $x = \begin{bmatrix} i_{s\alpha} & i_{s\beta} & \varphi_{r\alpha} & \varphi_{r\beta} \end{bmatrix}^T$, we obtain [39]

$$\dot{x}(t) = A(t)x(t) + B(t)u(t) \tag{26}$$

with:

$$A(t) = \begin{bmatrix} -\frac{1}{\sigma L_s}\left(R_s + \frac{1}{T_r}\frac{L_m^2}{L_r}\right) & 0 & \frac{1}{\sigma L_s}\left(\frac{L_m}{L_r}\frac{1}{T_r}\right) & \frac{1}{\sigma L_s}\left(\frac{L_m}{L_r}\right)\omega_r \\ 0 & -\frac{1}{\sigma L_s}\left(R_s + \frac{1}{T_r}\frac{L_m^2}{L_r}\right) & -\frac{1}{\sigma L_s}\left(\frac{L_m}{L_r}\right)\omega_r & \frac{1}{\sigma L_s}\left(\frac{L_m}{L_r}\frac{1}{T_r}\right) \\ \frac{L_m}{T_r} & 0 & -\frac{1}{T_r} & -\omega_r \\ 0 & \frac{L_m}{T_r} & \omega_r & -\frac{1}{T_r} \end{bmatrix} \quad (27)$$

$$B(t) = \begin{bmatrix} \frac{1}{L_s\sigma} & 0 \\ 0 & \frac{1}{L_s\sigma} \\ 0 & 0 \\ 0 & 0 \end{bmatrix} \quad (28)$$

$$\text{and} : u(t) = \begin{bmatrix} v_{s\alpha} \\ v_{s\beta} \end{bmatrix} \quad (29)$$

3.8 Simulation results

The purpose of this test is to validate our motor block before using it with space vector PWM (SVPWM) and with direct torque control. Our goal is to integrate it later in the simulations. To carry out the simulation, we translate the mathematical model of the machine using the SimPowerSystem blocks of the Matlab/Simulink software.

3.8.1 No-load test

For an induction machine supplied directly by the 220/380 V three-phase network and running off-load, we visualize the mechanical speed, the electromagnetic torque, the stator currents as well as the components of both the current and the stator fluxes.

The simulation results are represented in the **Figures 3–7**.

The steady-state speed stabilizes at a value close to the synchronism speed because the machine is not loaded. At no-load starting, the torque is strongly pulsating; it reaches a maximum value in the range of 3.2 times the nominal torque. This is due to the noises generated by the mechanical part, and after the disappearance of the transitory mode, it tends toward the value corresponding to the zero load. The absorbed current is high at start-up; it is about three times the rated. At steady state, there remains the current corresponding to the inductive behavior of the no-loaded motor. The rotor current is significant during start-up and drops completely at steady state.

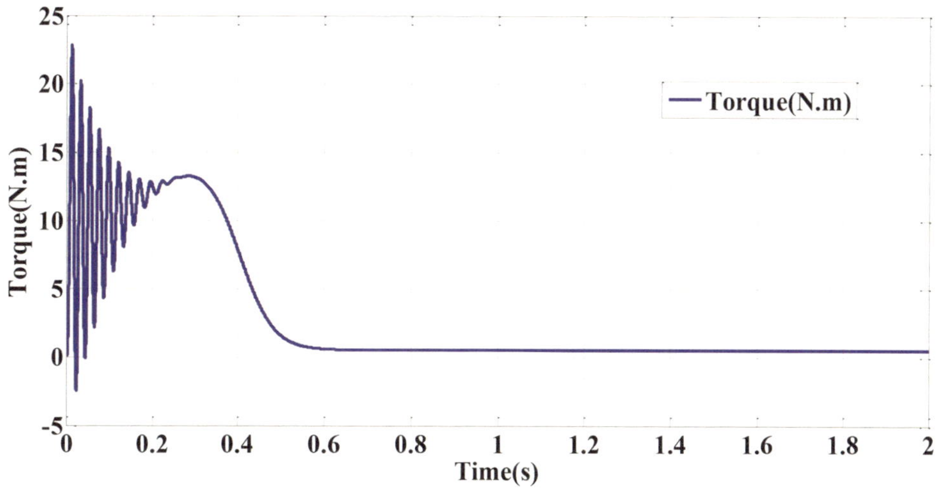

Figure 4.
Magnetic torque simulation result.

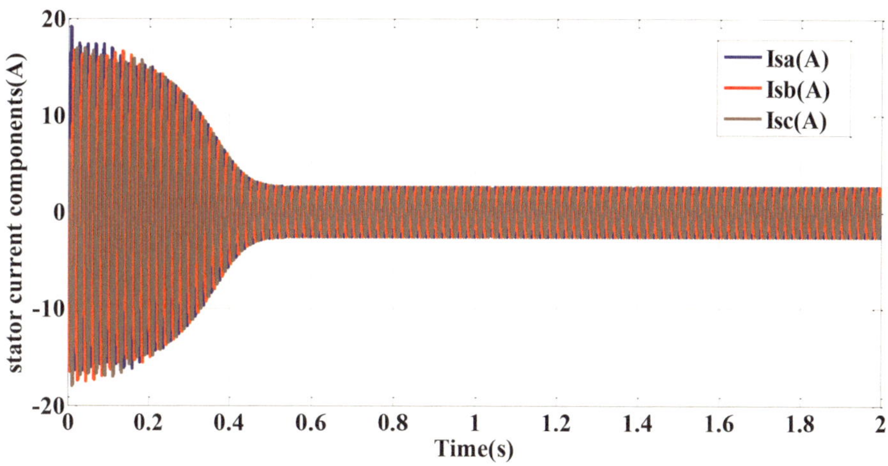

Figure 5.
Stator currents simulation result.

3.8.2 Load variations after no-load starting

Figures 8–12 represent the three-phase currents, the rotational speed, and the electromagnetic torque of the motor, respectively. Two cases are carried out in this simulation with no load and with a loaded motor:

When the motor run under no load condition, in the dynamic state, we observe an excessively absorbed current that stabilizes to produce a sinusoidal form with constant amplitude.

When the motor is not loaded, we observe at the beginning of the start-up running that the increase in speed is virtually linear, and the total inertia around the rotating shaft determines the speed-up time (about $0.5\,s$), where the obtained speed is close to $157\ rad/s$.

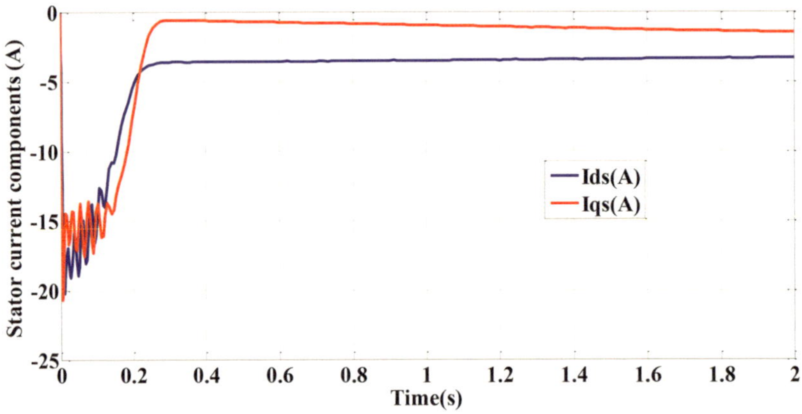

Figure 6.
Stator currents components simulation result.

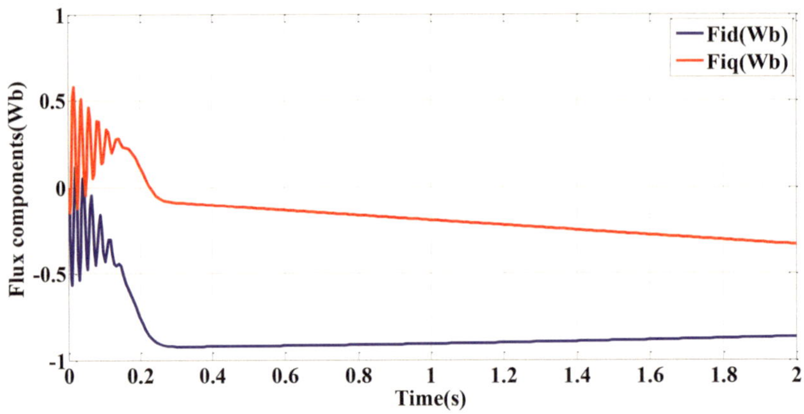

Figure 7.
Stator flux components simulation result.

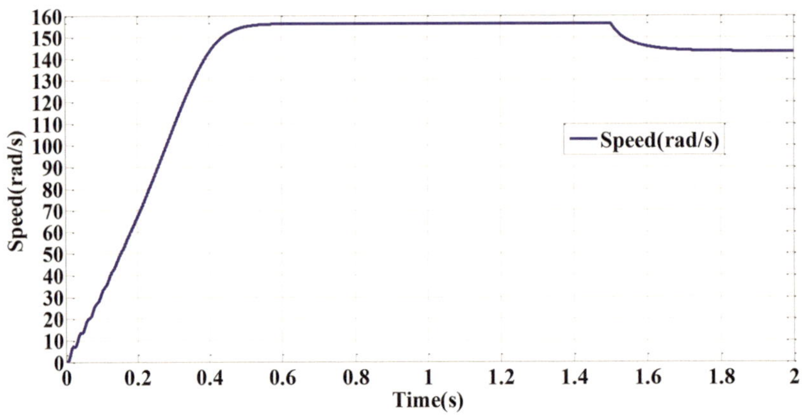

Figure 8.
Rotor speed simulation result.

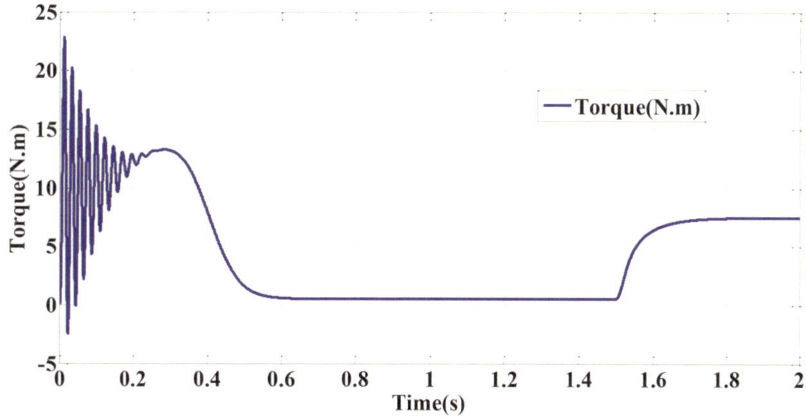

Figure 9.
Magnetic torque simulation result.

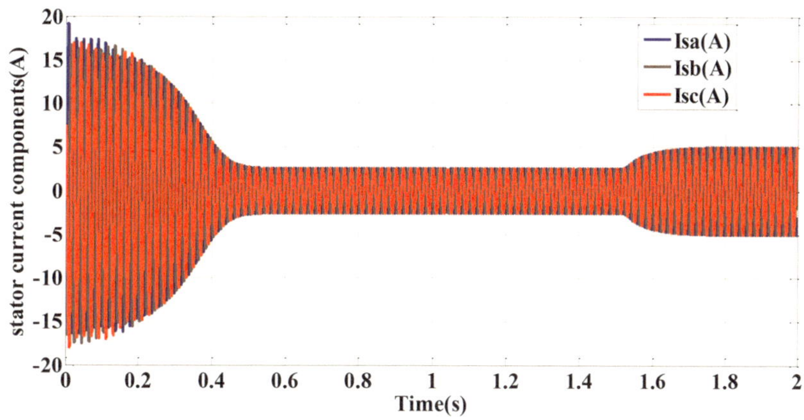

Figure 10.
Stator currents simulation result.

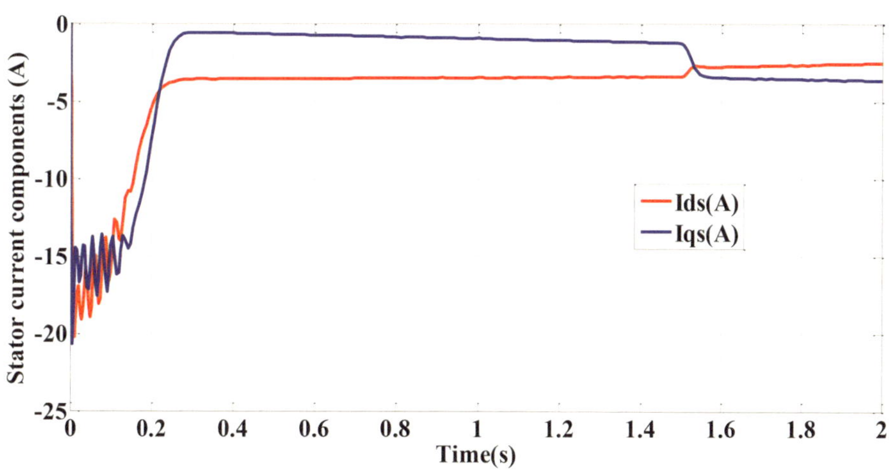

Figure 11.
Stator currents component simulation result.

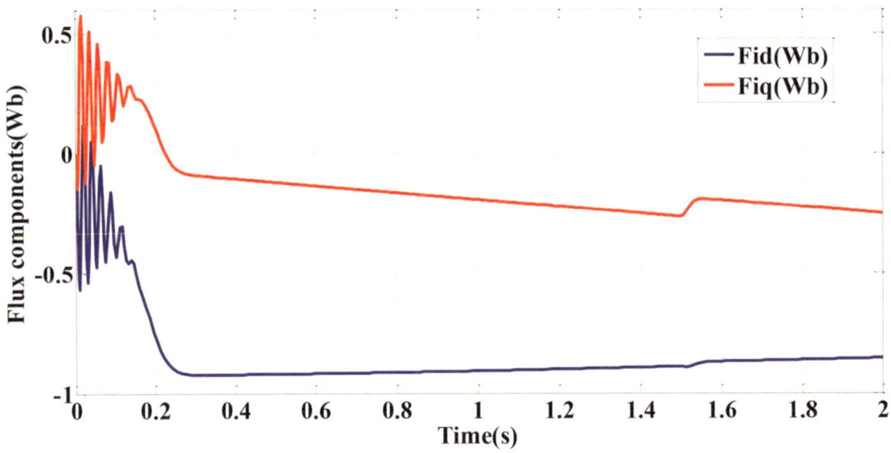

Figure 12.
Stator flux component simulation result.

Under load: A load torque ($\Gamma_r = 7\ N.m$) is applied to the machine shaft (at time $t = 1.5\ s$). When the electromagnetic torque reaches the load torque, obviously, there is a reduction in the rotating speed. Additionally, we notice an increase in the stator currents' magnitude and a slight decrease in the flux.

3.8.3 Starting under load

While starting under load (**Figures 13–17**), the electromagnetic torque responds instantly (because Γ_{em} is greater than Γ_r) and the asynchronous motor accelerates, where the speed is slightly disturbed. Without control, a high overshoot response for electromagnetic torque is obtained. Therefore, it is not recommended to be used in an open-loop system for stability reasons.

In steady-state operation, for the motor to operate correctly, the electromagnetic torque Γ_{em} must be equal to the resistive torque Γ_r. All of these characteristics and the moment of the resistive torque define the operating point of the induction machine.

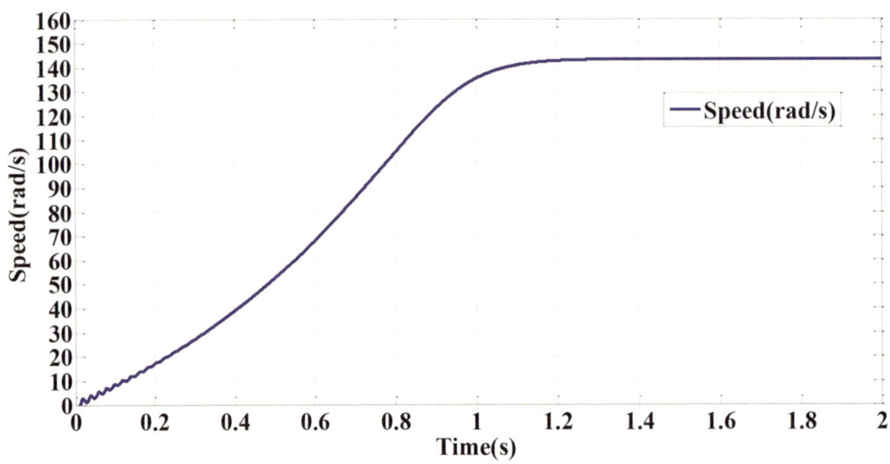

Figure 13.
Rotor speed simulation result.

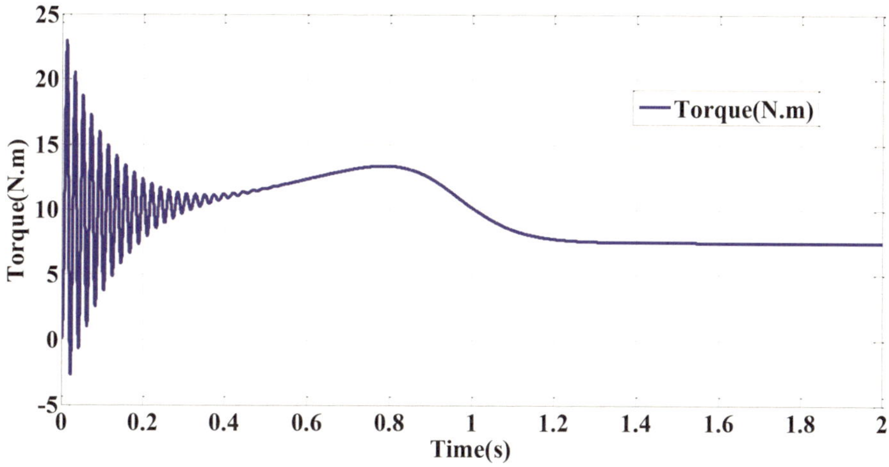

Figure 14.
Magnetic torque simulation result.

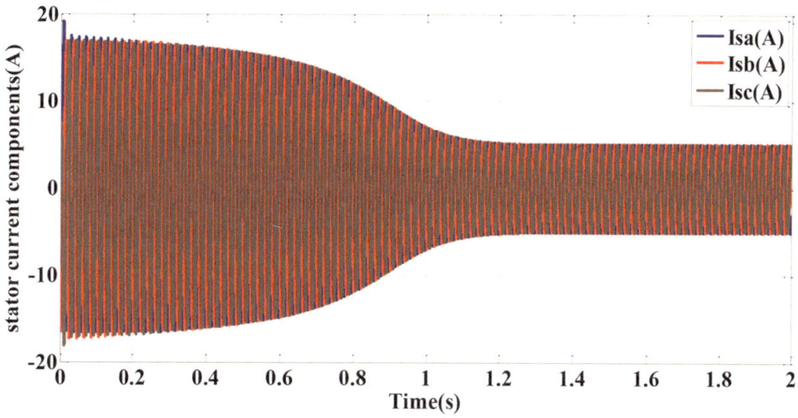

Figure 15.
Stator current simulation result.

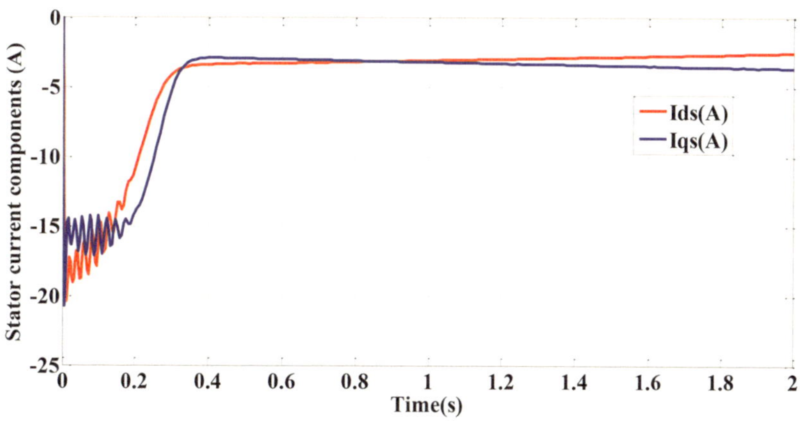

Figure 16.
Stator current component simulation result.

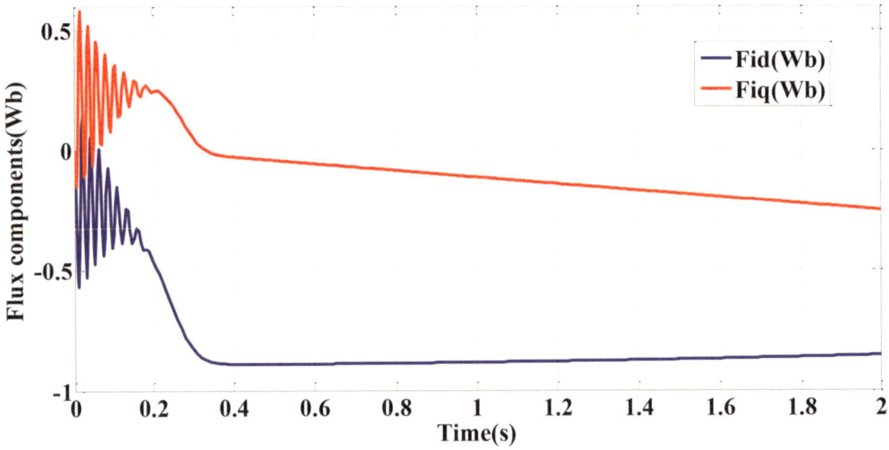

Figure 17.
Stator flux component simulation result.

4. Conclusion

In this chapter, a modeling analysis of induction motor has been done. First, we studied the modeling and open-loop simulation behavior of the squirrel cage induction motor. Its model is strongly nonlinear; however, by taking into account some simplification assumptions, the model become more simplified. The obtained simulation results show the validity of the model developed.

Author details

Abdelkarim Belbali[1]*, Salim Makhloufi[1], Abdellah Kadri[1], Laidi Abdallah[2] and Zemitte Seddik[2]

1 Faculty of Sciences and Technology, Energy, Environment and Computer Systems Laboratory (LESSI), Department of Sciences and Technology, Ahemd Draia University, Adrar, Algeria

2 Department of Science and Technology, Laboratory of Sustainable Development and Computer Science (LDDI), University of Adrar, Algeria

*Address all correspondence to: kar.belbali@univ-adrar.edu.dz

References

[1] Eldali F. A Comparitve Study between Vector Control and Direct Torque Control of Induction Motor Using MATLAB SIMULINK [Thesis]. USA: Colorado State University; 2012

[2] Farid TAZERART. Étude, Commande et Optimisation des Pertes d'Énergie d'une Machine à Induction Alimentée par un Convertisseur Matriciel [thesis]. Algeria: Bejaia University; 2016

[3] Shultz G. Transformers and Motors. Newnes; 1989

[4] Ozpineci B, Tolbert LM. Simulink implementation of induction machine model-a modular approach. In: IEEE International Electric Machines and Drives Conference, 2003. IEMDC'03. Vol. 2. IEEE; 2003. pp. 728-734

[5] Abdelkarim AMMAR. Amélioration des Performances de la Commande Directe de Couple (DTC) de La Machine Asynchrone par des Techniques Non-Linéaires [thesis]. Biskra: Université Mohamed Khider; 2017

[6] Available from: https://automationforum.co/three-phase-induction-motor-basics/.

[7] Caron JP, Hautier JP. Modélisation et commande de la machine asynchrone. Edition Technip ed1995

[8] Chatelain J. Machines électriques. Tome 1, Editions Dunod; 1987

[9] Mehazzem F. Contribution à la Commande d'un Moteur Asynchrone destiné à la Traction électrique [thesis]. Paris Est; 2010

[10] Grellet G, Clerc G. "Actionneurs électriques: Principes/Modèle/Commandes", Collection Electrotechnique. Edition Eyrolles; 1997

[11] Lesenne J, Notel F. "Introduction à l'électrotechnique approfondie", Technique et documentation. Paris, France; 1981

[12] Duan F, Zivanovic R. A model for induction motor with stator faults. In: 2012 22nd Australasian Universities Power Engineering Conference (AUPEC). IEEE; 2012. pp. 1-5

[13] Krause PC, Wasynczuk O, Sudhoff SD. Analysis of Electric Machinery. New York: IEEE Press; 1996

[14] Gandhi A, Corrigan T, Parsa L. Recent advances in modeling and online detection of stator interturn faults in electrical motors. IEEE Transactions on Industrial Electronics. 2010;**58**(5): 1564-1575

[15] AlShorman O, Irfan M, Saad N, Zhen D, Haider N, Glowacz A, et al. A review of artificial intelligence methods for condition monitoring and fault diagnosis of rolling element bearings for induction motor. Shock and vibration. 2020;**2020**

[16] Concordia C, Crary SB, Lyons JM. Stability characteristics of turbine generators. Electrical Engineering. 1938; **57**(12):732-744

[17] Tan G, Sun X, Xu W, Wang H, Li S. Power definition for three-phase unbalanced power system based on Tan-Sun coordinate transformation system. In: 2019 22nd International Conference on Electrical Machines and Systems (ICEMS). IEEE; 2019. pp. 1-6

[18] Nguyen NK, Kestelyn X, Dos Santos Moraes TJ. Generalized Vectorial

Formalism–Based Multiphase Series-Connected Motors Control, 2015

[19] Athari H, Niroomand M, Ataei M. Review and classification of control systems in grid-tied inverters. Renewable and Sustainable Energy Reviews. 2017;**72**:1167-1176

[20] Ali Z, Christofides N, Hadjidemetriou L, Kyriakides E, Yang Y, Blaabjerg F. Three-phase phase-locked loop synchronization algorithms for grid-connected renewable energy systems: A review. Renewable and Sustainable Energy Reviews. 2018;**90**: 434-452

[21] Li J, Konstantinou G, Wickramasinghe HR, Pou J. Operation and control methods of modular multilevel converters in unbalanced AC grids: A review. IEEE Journal of Emerging and Selected Topics in Power Electronics. 2018;**7**(2):1258-1271

[22] Abdennour D. Contrôle direct du couple du moteur à induction sans capteur de vitesse associée à un observateur non linéaire [thesis]. Algeria: Université de Batna;

[23] Yesma BENDAHA. Contribution à la commande avec et sans capteur mécanique d'un actionneur électrique [thesis]. UK: Université Mohamed Boudiaf des sciences et de la technologi; 2013

[24] Zerbo M. Identification des paramètres et commande vectorielle adaptative à orientation du flux rotorique de la machine asynchrone à cage [thesis]. Canada: Université du Québec à Trois-Rivières; 2008

[25] Jia H, Djilali N, Yu X, Chiang HD, Xie G. Computational science in smart grids and energy systems. Journal of Applied Mathematics. 2015;**2015**(2015): 326481

[26] Kundur P. Power System Stability and Control. New York, NY, USA: McGraw-Hill; 1993

[27] Cherifi D. Estimation de la vitesse et de la résistance rotorique pour la commande par orientation du flux rotorique d'un moteur asynchrone sans capteur mécanique. Université des Sciences et de la Technologie d'Oran Mohamed Boudiaf; 2014

[28] Lakhdar D. CONTRIBUTION a LA COMMANDE PREDICTIVE DIRECTE DU COUPLE DE LA MACHINE À INDUCTION [Thesis]. Université Mustapha Ben Boulaid Batna 2, Département de l'éléctrotechnique; 2017

[29] Aouragh N. Implémentation d'une Identification En Temps Réel De La Machine Asynchrone A Cage Sur Le DSP TMS320 LF 2404 A [thesis]. Université Mohamed Khider Biskra; 2005

[30] Rezgui SE, Benalla H. Commande de machine électrique en environnement Matlab/Simulink et temps réel [thesis]. Université Mentouri Constantine; 2009

[31] Robyns B, François B, Degobert P. Commande Vectorielle de la Machine Asynchrone: Désensibilisation et optimisation par la logique floue. Vol. 14. Editions TECHNIP; 2007

[32] Chatelain J. Machine électriques. tome I, Edition Dunod; 1983

[33] Eguiluz RP. Commande algorithmique d'un système mono-onduleur bimachine asynchrone destiné à la traction ferroviaire [thesis]. De l'INPT Toulouse; 2002

[34] Malatji MM. Derivation and Implementation of a DQ Model of an Induction Machine Using MATLAB/ SIMULINK

[35] Rezgui SE. Techniques de commande avancées de la machine asynchrone

[36] Yu J, Zhang T, Qian J. Electrical Motor Products: International Energy-Efficiency Standards and Testing Methods. Elsevier; 2011

[37] Thongam JS. Commande de Haute Performance Sans Capteur d'une Machine Asynchrone= High Performance Sensorless Induction Motor Drive. Université du Québec à Chicoutimi; 2006

[38] Khoury G. Energy Efficiency Improvement of a Squirrel-Cage Induction Motor through the Control Strategy [Thesis]. 2018

[39] Legrioui S, Benalla H. Observation et commande non-linéaire de la machine asynchrone avec identification on line des paramètres [thesis]. University of Mentouri Brothers Constantine;

[40] Bourbia W. Etude Comparée des Estimateurs de Vitesse pour la Commande de la Machine Asynchrone (THESE Présentée en vue de l'obtention du diplôme de DOCTORAT en sciences)

[41] Koteich M. Modélisation et observabilité des machines électriques en vue de la commande sans capteur mécanique [thesis]. Université Paris-Saclay; 2016

[42] Chaikhy H. Contribution au développement et à l'implantation des stratégies de commandes évoluées des machines asynchrones. 2013

[43] Meftah L. simulation et commande de la machine asynchrone double étoile pour aerogeneration. Mémoire de Magister, promotion; 2014

[44] Baghli L. Contribution à la commande de la machine asynchrone, utilisation de la logique floue, des réseaux de neurones et des algorithmes génétiques [thesis]. Université Henri Poincaré-Nancy I; 1999

[45] Kouzi K. Contribution des techniques de la logique floue pour la commande d'une machine à induction sans transducteur rotatif [thesis]. Université de Batna 2; 2008

[46] Duan F. Induction Motor Parameters Estimation and Faults Diagnosis Using Optimisation Algorithms [Thesis]

[47] Ghanes M. Observation et commande de la machine asynchrone sans capteur mécanique [thesis]. Ecole Centrale de Nantes (ECN); Université de Nantes; 2005

[48] Moussaoui L. Etude de la commande de l'ensemble machine asynchrone-Onduleur à source de courant [thesis]. Batna, Université El Hadj Lakhdar. Faculté des sciences de l'ingénieur; 2007

[49] Martins CDA. Contrôle direct du couple d'une machine asynchrone alimentée par convertisseur multiniveaux à fréquence imposée. 2000

[50] Camblong H. Minimisation de l'impact des perturbations d'origine éolienne dans la génération d'électricité par des aérogénérateurs à vitesse variable [thesis]. Bordeaux: Centre de Bordeaux; 2003

[51] Jamoussi K, Ouali M. et CHARRADI, Hassen. Reconstructeur d'état d'un moteur asynchrone par DSP. In: The Seventh International Conference on Science and Techniques of Automatic Control STA'2006. 2006. pp. 1-11

[52] Tarek BENMILOUD. Commande du moteur asynchrone avec compensation

des effets des variations paramétriques [thesis]. Université Mohamed Boudiaf des sciences et de la technologi; 2012

[53] Bouhafna S. Commande par DTC d'un moteur asynchrone apport des réseaux de neurones [thesis]. Université de Batna 2; 2013

Chapter 8

Model Reference Adaptive System Scheme for Sensorless Induction Motor Drive

Youssef Agrebi Zorgani, Soufien Hajji, Yessine Koubaa and Mohamed Boussak

Abstract

The induction machine is widely used in industrial fields due to its robustness, low cost, and good standardization. Indeed, its use in high-performance variable-speed drive systems requires the imposition of specific and complex control structures based on the mathematical model of the machine and its power supply. However, this chapter book deals with indirect stator field-oriented control (ISFOC) of an induction motor drive (IM), without a speed sensor. In previous works, the MRAS scheme has been used to estimate the speed and the rotor resistance. The reference model and the adjustable one, which are developed in the stationary stator reference frame, are used to estimate simultaneously the rotor speed and the rotor resistance (IM) from the knowledge of the stator currents and voltages. Simulation and experimental results are presented to validate the mathematical study as well as to prove the effectiveness and the robustness of the proposed scheme of control and sensorless ISFOC induction motor drive.

Keywords: induction machine, stator flux orientation model, model reference adaptive system (MRAS), rotor speed, resistance estimation

1. Introduction

The induction motor IM has long been used mainly at a constant speed. Its control remains a challenge for researchers, in order to optimize and control the motor in variable-speed drives. With advances in power electronics such as the appearance of GTO thyristors and subsequently IGBT transistors as well as digital electronics such as the development of new DSPs (Digital Signal Processors), the drive problem at variable speed is resolved and the control strategies have been implemented under satisfactory conditions.

The vector control (FOC: Field-Oriented Control) also called flux orientation control, it allows to reduce the behavior of the IM to that of a direct current DC motor (with separate excitation) to have the same performance in torque and speed [1, 2]. The FOC has very good accuracy for torque and speed. Moreover, this strategy, although sophisticated, has the disadvantage of requiring the installation of a mechanical sensor on the shaft of the motor and also has a high sensitivity to the parametric variations of the IM in particular to the rotor resistance [3]. Two methods are possible for flux-oriented vector control [4]. The first is called the Direct Control Method (DFOC) and the second is called Indirect Method (IFOC). This strategy consists of not estimating the flux but directly using the amplitude of its reference value ϕ^*.

To address the problem of the presence of a speed sensor in the IFOC, we opt for a control without a mechanical sensor, which is based on the design of software sensors for the estimation of physical variables, such as the speed and rotor resistance. The design of such sensors is mainly based on the synthesis of observers or adaptive methods allowing parametric identification and sensorless control of the IM. In the literature, several methods of simultaneous estimation of speed and rotor resistance in sensorless IM drive have appeared. Among the most used techniques, we can mention:

- Rotor resistance estimation using the transient state under the speed sensorless control of the induction motor [5].

- Speed sensorless of IM drive with online model parameter tuning for steady-state accuracy [6, 7].

- Simultaneous estimation of the rotor speed and the rotor resistance was estimated by adding a small alternating current to the rotor flux [8].

This chapter book presents a modeling and control of an asynchronous machine with and without mechanical sensors. However, the measurement of the currents and voltages will be used in order to estimate the speed and the rotor resistance using the Model Reference Adaptive System (MRAS) scheme. A theoretical study is given with detailed mathematical equations and accompanied by simulation and experimental results in order to demonstrate the effectiveness of the proposed method.

This chapter book is organized as follows. Section 2 presents an asynchronous motor model. While, in Section 3, an induction motor state model will be introduced. Section 4 formulates the stator field orientation control of the induction motor drive. Then, the simultaneous estimation of speed and rotor resistance will be illustrated and discussed in the section. Furthermore, the simultaneous estimation of speed and rotor resistance using MRAS technique. Simulation results are proposed in Section 6. Thereafter, in Section 7, we will present the experimental results obtained on a test bench provided with a chart of real-time control of the type dS1104. Finally, we ended up with a conclusion.

2. Asynchronous machine model

According to the laws of physics, the asynchronous machine can be associated with mathematical equations that involve its parameters. The implementation of these equations gives rise to the modeling. At the end of this operation, a problem arises; however, when the model is closest to reality, it becomes very complex and requires a very powerful and high-performance calculator. In order to simplify the calculation while keeping the most important phenomena and neglecting the secondary phenomena, and for the resulting model to be usable in both static and dynamic conditions, we must use simplifying assumptions.

2.1 Simplifying assumptions

The equations of the asynchronous machine are made by adopting the following simplifying assumptions [9, 10]:

- The air gap is assumed to be constant, and the machine is symmetrical.
- The current density is uniform in the conductor sections.
- The magnetic circuit is unsaturated and perfectly laminated to the stator and to the rotor.
- Winding resistances do not vary with temperature and skin and notch effects are neglected.
- The spatial distribution of magnetomotor forces is assumed to be sinusoidal along the air gap.
- The cage rotor is described as a balanced three-phase winding.

2.2 Mathematical model of the asynchronous machine

According to **Figure 1**, the Induction Motor (IM) has six windings:

- The stator consists of three fixed windings (A,B,C) offset in the space of 120° from each other and crossed by three sinusoidal currents.
- The rotor can be modeled by three identical windings and shorted (a,b,c) offset by 120° in space.

2.2.1 Electrical equation

The Ohm law applied to the stator and rotor circuits is written in the following matrix form:

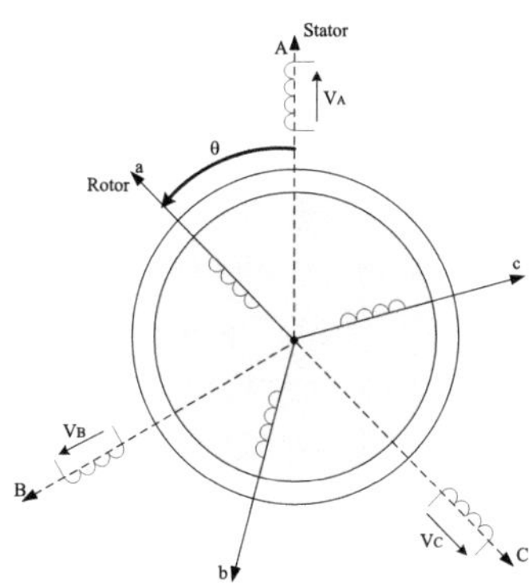

Figure 1.
Spatial representation of asynchronous machine windings.

$$\begin{bmatrix} v_{as} \\ v_{bs} \\ v_{cs} \end{bmatrix} = [R_s] \begin{bmatrix} i_{as} \\ i_{bs} \\ i_{cs} \end{bmatrix} + \frac{d}{dt} \begin{bmatrix} \phi_{as} \\ \phi_{bs} \\ \phi_{cs} \end{bmatrix} \tag{1}$$

$$\begin{bmatrix} v_{ar} \\ v_{br} \\ v_{cr} \end{bmatrix} = [R_r] \begin{bmatrix} i_{ar} \\ i_{br} \\ i_{cr} \end{bmatrix} + \frac{d}{dt} \begin{bmatrix} \phi_{ar} \\ \phi_{br} \\ \phi_{cr} \end{bmatrix} \tag{2}$$

The matrices of the asynchronous machine stator and rotor resistors are given by:

$$R_s = \begin{bmatrix} R_s & 0 & 0 \\ 0 & R_s & 0 \\ 0 & 0 & R_s \end{bmatrix} \text{ and } R_r = \begin{bmatrix} R_r & 0 & 0 \\ 0 & R_r & 0 \\ 0 & 0 & R_r \end{bmatrix} \tag{3}$$

2.2.2 Magnetic equation

According to the simplifying assumptions, the relations between fluxes and currents are linear and can be written in the following matrix form:

$$\begin{bmatrix} \phi_{as} \\ \phi_{bs} \\ \phi_{cs} \end{bmatrix} = [L_{ss}] \begin{bmatrix} i_{as} \\ i_{bs} \\ i_{cs} \end{bmatrix} + [L_{sr}] \begin{bmatrix} i_{ar} \\ i_{br} \\ i_{cr} \end{bmatrix} \tag{4}$$

$$\begin{bmatrix} \phi_{ar} \\ \phi_{br} \\ \phi_{cr} \end{bmatrix} = [L_{rr}] \begin{bmatrix} i_{ar} \\ i_{br} \\ i_{cr} \end{bmatrix} + [L_{sr}] \begin{bmatrix} i_{as} \\ i_{bs} \\ i_{cs} \end{bmatrix} \tag{5}$$

$$\text{With : } [L_{ss}] = \begin{bmatrix} L_{as} & M_{as} & M_{as} \\ M_{as} & L_{as} & M_{as} \\ M_{as} & M_{as} & L_{as} \end{bmatrix} ; [L_{rr}] = \begin{bmatrix} L_{ar} & M_{ar} & M_{ar} \\ M_{ar} & L_{ar} & M_{ar} \\ M_{ar} & M_{ar} & L_{ar} \end{bmatrix} \tag{6}$$

$$\text{and } [L_{sr}] = \begin{bmatrix} \cos(\theta) & \cos\left(\theta + \frac{2\pi}{3}\right) & \cos\left(\theta - \frac{2\pi}{3}\right) \\ \cos\left(\theta - \frac{2\pi}{3}\right) & \cos(\theta) & \cos\left(\theta + \frac{2\pi}{3}\right) \\ \cos\left(\theta + \frac{2\pi}{3}\right) & \cos\left(\theta - \frac{2\pi}{3}\right) & \cos(\theta) \end{bmatrix} \tag{7}$$

We finally obtain the three-phase asynchronous model:

$$\begin{cases} [v_{sabc}] = [R_s] \cdot [i_{sabc}] + \frac{d}{dt}\{[L_s] \cdot [i_{sabc}] + [L_{sr}] \cdot [i_{rabc}]\} \\ [v_{rabc}] = [R_r] \cdot [i_{rabc}] + \frac{d}{dt}\{[L_r] \cdot [i_{rabc}] + [L_{rs}] \cdot [i_{sabc}]\} \end{cases} \tag{8}$$

with:

$[v_{sabc}] = [v_a \ v_b \ v_c]_s^T$, $[v_{rabc}] = [v_a \ v_b \ v_c]_r^T$, $[i_{sabc}] = [i_a \ i_b \ i_c]_s^T$ and $[i_{rabc}] = [i_a \ i_b \ i_c]_r^T$

2.2.3 Mechanical equation

The fundamental relationship of dynamics allows us to write:

$$T_{em} - T_l = J\frac{d\Omega}{dt} + f \cdot \Omega \tag{9}$$

2.3 Two-phase model of the machine

When studying transient phenomena, we are faced with the following problem:

The Eq. (6) of the matrix of mutual inductances being with nonconstant elements and the coefficients of the Eq. (8) are variable from where the analytical resolution of this system of equations becomes very difficult. To tackle this problem, we use mathematical transformations to describe the behavior of the machine using differential equations with constant coefficients, which facilitate its resolution. These transformations must preserve the instantaneous power and the reciprocity of mutual inductances.

The use of Park's transformation makes it possible to circumvent these problems and simplify the statoric and rotoric equations in order to obtain a system of equations with constant coefficients, which considerably reduces the calculation time.

2.3.1 Park transformation

The transformation of Park makes it possible to go from a three-phase system abc to a two-phase system d and q axes. A Park P(β) matrix allows the passage of the X_{abc} quantities from the three-phase system to the two-phase X_{dq} quantities, rotating at a speed that depends on the stator or rotor quantities, such as (**Figure 2**).

$$[P(\beta)] = \sqrt{\frac{2}{3}}\begin{bmatrix} \cos(\beta) & \cos\left(\beta - \frac{2\pi}{3}\right) & \cos\left(\beta + \frac{2\pi}{3}\right) \\ -\sin(\beta) & -\sin\left(\beta - \frac{2\pi}{3}\right) & -\sin\left(\beta + \frac{2\pi}{3}\right) \\ \frac{1}{\sqrt{2}} & \frac{1}{\sqrt{2}} & \frac{1}{\sqrt{2}} \end{bmatrix} \tag{10}$$

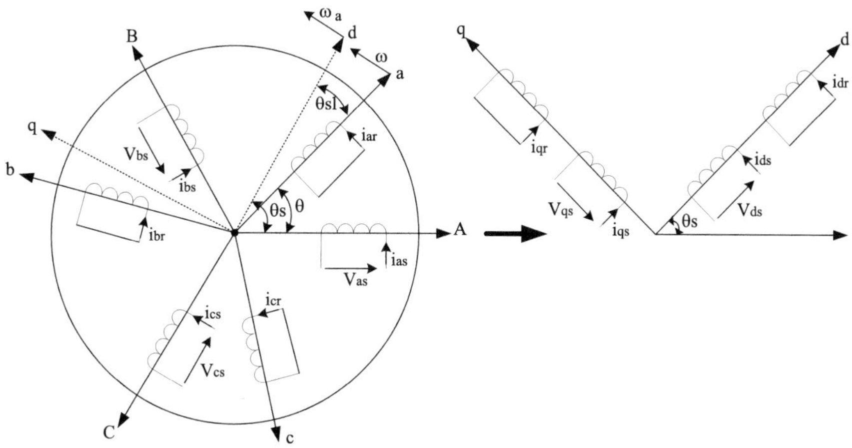

Figure 2.
Representation of the equivalent three-phase and two-phase asynchronous machine.

$$\begin{bmatrix} X_d \\ X_q \end{bmatrix} = [P(\beta)]^{-1} \cdot \begin{bmatrix} X_a \\ X_b \\ X_c \end{bmatrix} \tag{11}$$

This matrix ensures the instantaneous power invariance and its calculation is obtained by taking $\beta = \theta_s$, for the stator quantities and $\beta = \theta_{sl}$ for the rotoric quantities.

2.3.2 Electrical model of the asynchronous machine

By applying the Park transformation to the electric and magnetic equations of the asynchronous machine, we obtain the Eqs. (12) and (13), which are expressed in a reference frame running at an arbitrary speed ωa.

Stator equations.

$$\begin{cases} V_{ds} = R_s i_{ds} + \dfrac{d\phi_{ds}}{dt} - \omega_a \phi_{qs} \\ V_{qs} = R_s i_{qs} + \dfrac{d\phi_{qs}}{dt} + \omega_a \phi_{ds} \\ V_{0s} = R_s i_{0s} + \dfrac{d\phi_{0s}}{dt} \end{cases} \tag{12}$$

Rotor equations.

$$\begin{cases} V_{dr} = R_r i_{dr} + \dfrac{d\phi_{dr}}{dt} - \omega_{sl} \phi_{qr} \\ V_{qr} = R_r i_{qr} + \dfrac{d\phi_{qr}}{dt} + \omega_{sl} \phi_{dr} \\ V_{0r} = R_r i_{0r} + \dfrac{d\phi_{0r}}{dt} \end{cases} \tag{13}$$

with

$$\begin{bmatrix} \phi_{ds} \\ \phi_{qs} \\ \phi_{dr} \\ \phi_{qr} \end{bmatrix} = \begin{bmatrix} L_s & 0 & M & 0 \\ 0 & L_s & 0 & M \\ M & 0 & L_r & 0 \\ 0 & M & 0 & L_r \end{bmatrix} \begin{bmatrix} i_{ds} \\ i_{qs} \\ i_{dr} \\ i_{qr} \end{bmatrix} \tag{14}$$

For a balanced system, the stator and rotor homopolar components are zero. Knowing that the machine is caged and taking into account Eqs. (13), the Eqs. (11) and (12) are reduced to:

$$\begin{cases} V_{ds} = R_s i_{ds} + L_s \dfrac{di_{ds}}{dt} + M \dfrac{di_{dr}}{dt} - \omega_a (L_s i_{qs} + M i_{qr}) \\ V_{qs} = R_s i_{qs} + L_s \dfrac{di_{qs}}{dt} + M \dfrac{di_{qr}}{dt} + \omega_a (L_s i_{ds} + M i_{dr}) \\ 0 = R_r i_{dr} + L_r \dfrac{di_{dr}}{dt} + M \dfrac{di_{ds}}{dt} - \omega_{sl} (L_r i_{qr} + M i_{qs}) \\ 0 = R_r i_{qr} + L_r \dfrac{di_{qr}}{dt} + M \dfrac{di_{qs}}{dt} + \omega_{sl} (L_r i_{dr} + M i_{ds}) \end{cases} \tag{15}$$

2.3.3 Electromagnetic couple expressions

We can see that the torque results from an interaction between flows and currents. The application of the transformation of Park provides us with several forms of expressions of the electromagnetic couple among which we distinguish:

$$T_e = n_p \frac{M}{L_r} \left(\phi_{dr} i_{qs} - \phi_{qr} i_{ds} \right) \tag{16}$$

$$T_e = n_p \left(\phi_{ds} i_{qs} - \phi_{qs} i_{ds} \right) \tag{17}$$

The choice of the expression will depend on the type of fluxes orientation. In fact, we use (16) for the rotor orientation and (17) for the stator flux orientation.

3. Induction motor state model

The state model of an asynchronous machine allows the study of its transient states. Eq. (15), which forms the state equations of the machine, represents a nonlinear differential system with constant coefficients. ($\omega_a = \omega_s$ and $\omega_{sl} = \omega_s - \omega$).

The possibilities for choosing electrical state variables are diverse. This choice is characterized by the variables endowed with memory, such as the currents, the fluxes, the speed, and the angular position.

The machine state model is:

$$\begin{cases} \left[\dfrac{dX}{dt}\right] = [A][X] + [B][U] \\ [Y] = [C][X] \end{cases} \tag{18}$$

The quantities of the state vector are chosen according to the quantities to be controlled.

3.1 State model for stator flux orientation control

We consider the second expression of the electromagnetic couple given by Eq. (18).

The state variables used are ϕ_{ds}, ϕ_{qs}, $i_{ds,}$ and i_{qs}. For this reason, we are interested to eliminate the flux ϕ_{dr}, $\phi_{qr,}$ and the current i_{dr} and i_{qr} of the Eq. (15). This leads to a state representation of the asynchronous machine of the form:

$$\left[\frac{dX}{dt}\right] = [A][X] + [B][U] \tag{19}$$

with:

$$X = \begin{bmatrix} \phi_{ds} \\ \phi_{qs} \\ i_{ds} \\ i_{qs} \end{bmatrix}^T ; U = \begin{bmatrix} v_{ds} \\ v_{qs} \end{bmatrix}^T ; B = \begin{bmatrix} 1 & 0 \\ 0 & 1 \\ \dfrac{1}{\sigma L_s} & 0 \\ 0 & \dfrac{1}{\sigma L_s} \end{bmatrix}.$$

$$A = \begin{bmatrix} 0 & \omega_s & -R_s & 0 \\ -\omega_s & 0 & 0 & -R_s \\ \frac{R_r}{\sigma L_s L_r} & \frac{\omega}{\sigma L_s} & -\frac{1}{\sigma L_s}\left(R_s + \frac{R_r L_s}{L_r}\right) & \omega_{sl} \\ -\frac{\omega}{\sigma L_s} & \frac{R_r}{\sigma L_s L_r} & -\omega_{sl} & -\frac{1}{\sigma L_s}\left(R_s + \frac{R_r L_s}{L_r}\right) \end{bmatrix}$$

3.2 State model for rotor flux orientation control

We consider the second expression of the electromagnetic couple given by Eq. (18). The state variables used are ϕ_{dr}, ϕ_{qr}, i_{ds}, and i_{qs}. For this reason, we are interested to eliminate the flux ϕ_{ds}, ϕ_{qs} and the current i_{dr} and i_{qr} of the Eq. (15). This leads to a state representation of the asynchronous machine of the form:

$$X = \begin{bmatrix} \phi_{dr} \\ \phi_{qr} \\ i_{ds} \\ i_{qs} \end{bmatrix}^T ; U = \begin{bmatrix} v_{ds} \\ v_{qs} \end{bmatrix}^T ; B = \begin{bmatrix} 1 & 0 \\ 0 & 1 \\ \frac{1}{\sigma L_s} & 0 \\ 0 & \frac{1}{\sigma L_s} \end{bmatrix} A$$

$$= \begin{bmatrix} -\frac{R_r}{L_r} & (\omega_s - \omega) & \frac{MR_r}{L_r} & 0 \\ -(\omega_s - \omega) & -\frac{R_r}{L_r} & 0 & \frac{MR_r}{L_r} \\ \frac{R_r M^2}{\sigma L_s L_r^2} & \frac{M}{\sigma L_s L_r}\omega & -\frac{1}{\sigma L_s}\left(R_s + \frac{R_r M^2}{L_r^2}\right) & \omega_s \\ -\frac{M}{\sigma L_s L_r}\omega & \frac{R_r M^2}{\sigma L_s L_r^2} & -\omega_s & -\frac{1}{\sigma L_s}\left(R_s + \frac{R_r M^2}{L_r^2}\right) \end{bmatrix}.$$

3.3 Simulation

The model of the asynchronous machine was tested by simulation with the Matlab-Simulink software. **Figure 3(a)** represents the curve of the speed during the direct start of the induction machine on the network with an application and cancelation of a load at time t = 0.4 s and at t = 0.7 s, respectively.

The analysis of the previous results shows that the IM undergoes a very brutal transient state during its direct start on the network. The method for alleviating the various constraints consists in supplying the motor with a set of variable voltages in frequency and amplitude. Various approaches are possible and they are all based on the steady-state characteristics of the induction motor.

4. Stator field-oriented control of induction motor drive

Vector control offers a better solution to achieve better performance in variable-speed applications. This solution appeared with the work of Blaschke in early 1970.

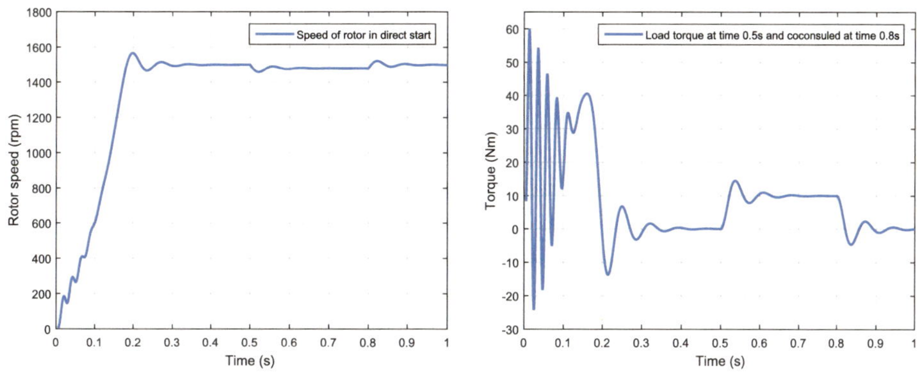

(a) (b)

Figure 3.
Speed and torque of an asynchronous machine during a direct start.

The vector control also called the flux orientation controller, consists in orienting the flux according to axis d, and therefore its component in axis quadrature q is zero in order to make the behavior of the asynchronous machine similar to that of a direct-current machine with independent excitation. The purpose of this control is to eliminate the coupling between the armature and the inductor, so that its operation is comparable to that of a direct-current machine, by breaking down the stator current into two components, one controlling the flow and the other controlling the torque. In our work, we will focus on the control by direction of the statoric flux.

By referring to a rotating reference frame, denoted by the superscript (d,q), the dynamic model of a three-phase induction motor can be expressed as follows [9, 10]:

$$\begin{bmatrix} \frac{d\phi_{ds}}{dt} \\ \frac{d\phi_{qs}}{dt} \\ \frac{di_{ds}}{dt} \\ \frac{di_{qs}}{dt} \\ \frac{d\omega}{dt} \end{bmatrix} = \begin{bmatrix} 0 & \omega_s & -R_s & 0 & 0 \\ -\omega_s & 0 & 0 & -R_s & 0 \\ \frac{1}{\sigma\tau_r L_s} & \frac{\omega}{\sigma L_s} & -\frac{1}{\sigma}\left(\frac{\tau_s+\tau_r}{\tau_s\tau_r}\right) & \omega_{sl} & 0 \\ -\frac{\omega_r}{\sigma L_s} & \frac{1}{\sigma\tau_r L_s} & -\omega_{sl} & -\frac{1}{\sigma}\left(\frac{\tau_s+\tau_r}{\tau_s\tau_r}\right) & 0 \\ -\frac{N_p^2}{J}i_{qs} & \frac{N_p^2}{J}i_{ds} & 0 & 0 & -\frac{f}{J} \end{bmatrix} \begin{bmatrix} \phi_{ds} \\ \phi_{qs} \\ i_{ds} \\ i_{qs} \\ \omega \end{bmatrix} + \begin{bmatrix} 1 & 0 & 0 \\ 0 & 1 & 0 \\ \frac{1}{\sigma L_s} & 0 & 0 \\ 0 & \frac{1}{\sigma L_s} & 0 \\ 0 & 0 & -\frac{1}{J} \end{bmatrix} \begin{bmatrix} v_{ds} \\ v_{qs} \\ N_p T_l \end{bmatrix} \quad (20)$$

with $\omega_{sl} = \omega_s - \omega$, $\tau_r = \frac{L_r}{R_r}$, $\tau_s = \frac{L_s}{R_s}$ and $\sigma = 1 - \frac{M^2}{L_s L_r}$.

The torque (Eq. (17)) looks like a DC machine if the second product is eliminated $\phi_{qs}I_{ds}$. So, it is enough to orient the reference (d,q), so as to cancel the quadrature flux component, in other words, to orient the stator flux according to the axis d ($\phi_{ds} = \phi_s$). Therefore, its quadrature component of axis q is zero ($\phi_{qs} = 0$). In this case, the stator flux will be controlled by the d-axis stator current and the torque is controlled by the q-axis quadrature current. The new expression of the couple is: $T_e = n_p\left(\phi_{ds} i_{qs}\right)$.

Using Eq. (20), the d and q-axis command stator currents are presented as:

$$i_{ds}^* = \frac{(1+\tau_r p)\Phi_s^* + L_s \sigma \tau_r \omega_{sl}^* i_{qs}^*}{L_s(1+\sigma\tau_r p)} \tag{21}$$

$$i_{qs}^* = \frac{\tau_r \omega_{sl}^* \left(\Phi_s^* - L_s \sigma i_{ds}^*\right)}{L_s(1+\sigma\tau_r p)} \tag{22}$$

The reference d and q-axis command voltages are expressed by the following equations:

$$v_{ds}^* = \sigma L_s \left(p + \frac{\tau_s + \tau_r}{\sigma \tau_s \tau_r}\right) i_{ds}^* - \frac{\Phi_s^*}{\tau_r} - \sigma L_s \omega_{sl}^* i_{qs}^* \tag{23}$$

$$v_{qs}^* = \sigma L_s \left(p + \frac{\tau_s + \tau_r}{\sigma \tau_s \tau_r}\right) i_{qs}^* + \frac{\omega_r}{\tau_r} \Phi_s^* + \sigma L_s \omega_{sl}^* i_{ds}^* \tag{24}$$

The IP speed controller is designed in order to stabilize the closed-loop sensorless speed control. The closed-loop transfer function is given by:

$$\frac{\omega(p)}{\omega^*(p)} = \frac{\omega_{nv}}{p^2 + 2\xi_v \omega_{nv} p + \omega_{nv}^2} \tag{25}$$

With: $\omega_{nv}^2 = \frac{N_p K_i K_p \lambda}{J}$, $2\xi_v \omega_{nv} = \frac{f + N_p K_p \lambda}{J}$, and $\lambda = \frac{L_s}{N_p \Phi_s^2 \tau_r (1-\sigma)}$.

The gains of the IP controller, K_p and K_i, are determined using a design method to obtain a damping ratio of 1. The gains parameters values of the IP speed controller are:

$$K_{vp} = R_r \alpha,\ \alpha = \frac{J - \tau_2 \Phi_s}{N_p \lambda \tau_2},\ \text{and}\ \lambda = \frac{L_s}{N_p \Phi_s L_s (1-\sigma)}$$

$$K_{vi} = \frac{N_p \beta + f}{4 N_p \beta \tau_2},\ \text{and}\ \beta = \frac{J - \tau_2 f}{N_p \tau_2},\ \tau_2 = \frac{J\omega}{n_p T_{eMax}}$$

For d- and q-axis current controller, we used a proportional-integral (PI) controller. The closed-loop transfer function is given by:

$$\frac{i_{ds}(p)}{i_{ds}^*(p)} = \frac{i_{qs}(p)}{i_{qs}^*(p)} = \frac{\omega_{ni}^2}{p^2 + 2\xi_i \omega_{ni} p + \omega_{ni}^2}\left(\frac{K_{ip}}{K_{ii}} p + 1\right) \tag{26}$$

With: $\omega_{ni}^2 = \frac{K_c K_{ii}}{\tau_c}$, $2\xi_i \omega_n = \frac{1 + K_c K_{ip}}{\tau_c} K_c = \frac{\tau_r}{R_s(\tau_s + \tau_r)}$, and $\tau_c = \frac{\sigma \tau_s \tau_r}{\tau_s + \tau_r}$.

Using the same method as the speed controller, the gains K_{ip} and K_{ii} parameters values of the PI current controller are:

$$K_{ii} = \left(\frac{\sigma L_s}{\tau_1{}^2}\right), K_{ip} = \rho - R_r \frac{L_s}{L_r} \text{ and } \rho = \frac{2\sigma L_s}{\tau_1} - R_s.$$

In our simulation, we have used the time constant $\tau_1 = 2ms$. We use a pulse-wide modulation (PWM) voltage source inverter.

4.1 Simulation results

Figure 4 represents the block diagram of the control by stator flux orientation of a 3 kW asynchronous machine with a mechanical sensor. It consists of a three-phase cage asynchronous machine powered by a pulse-width modulation (PWM) voltage inverter. A block for regulating the rotor speed with an IP corrector and that of the current with a PI corrector.

The ISFOC of an asynchronous machine with a mechanical sensor was tested by simulation with the Matlab-Simulink software. In this study, we regulate the speed with an IP corrector (Kvi = 7.9665, Kvp = 0.3001) and the current by a PI corrector (Kip = 26.929, Kii = 7764.8). The set points are the stator flux and the speed.

4.1.1 Influence of load torque variation

In this part, we simulated the system for a speed setpoint of 1000 rpm at time 1 s, under the application of a load torque equal to 10 Nm between times t_1 = 3 s and t_2 = 5 s. We obtained the following results:

Figure 5(a) represents the curve of the speed during a load application at time t = 3 s and its cancelation at time t = 5 s, we can clearly see that the real speed is stable and converges toward its reference value, which shows the reliability of the proposed

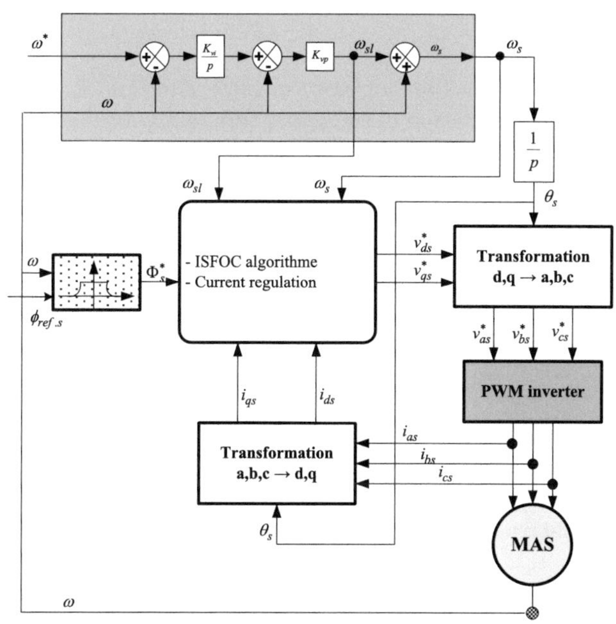

Figure 4.
Block diagram of the ISFOC of an asynchronous machine with speed sensor.

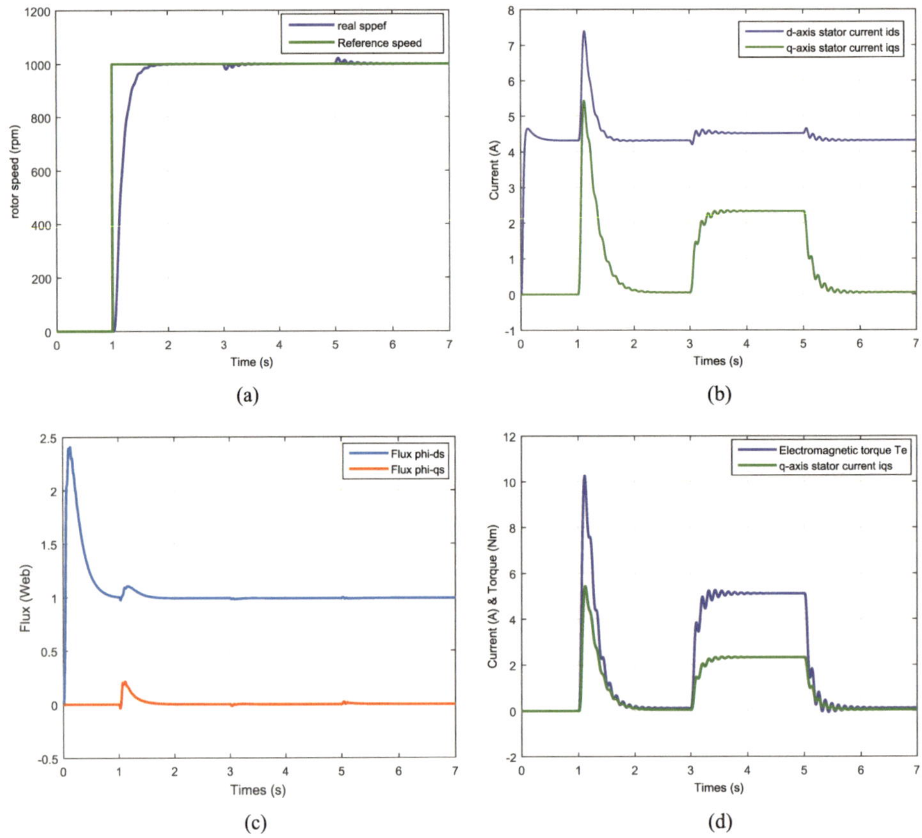

Figure 5.
Indirect stator flux-oriented controlled (ISFOC) induction motor drive under load torque.

algorithm. **Figure 5(b)** shows the curve of the current ids and iqs. However, **Figure 5(c)** proves the orientation of the stator flux that we have chosen, so we notice that $\phi_{ds} = \phi_s$ and $\phi_{qs} = 0$. Furthermore, **Figure 5(d)** shows that the curve of the electromagnetic torque and the quadratic current iqs are proportional, even during a load variation.

4.1.2 Low-speed study and inversion of the direction of rotation

To increase the robustness of the control, a low speed was applied with a rotation reversal of [0, 20, 0, −20, 0] rpm at times 1 s, 3 s, 4 s, and 6 s. The simulation results are presented in **Figure 6**:

5. Simultaneous estimation of the rotor speed and the rotor resistance

5.1 Description of the MRAS technique

The principle of estimation by this technique is based on the comparison of the outputs of two estimators from the knowledge of currents and stator voltages.

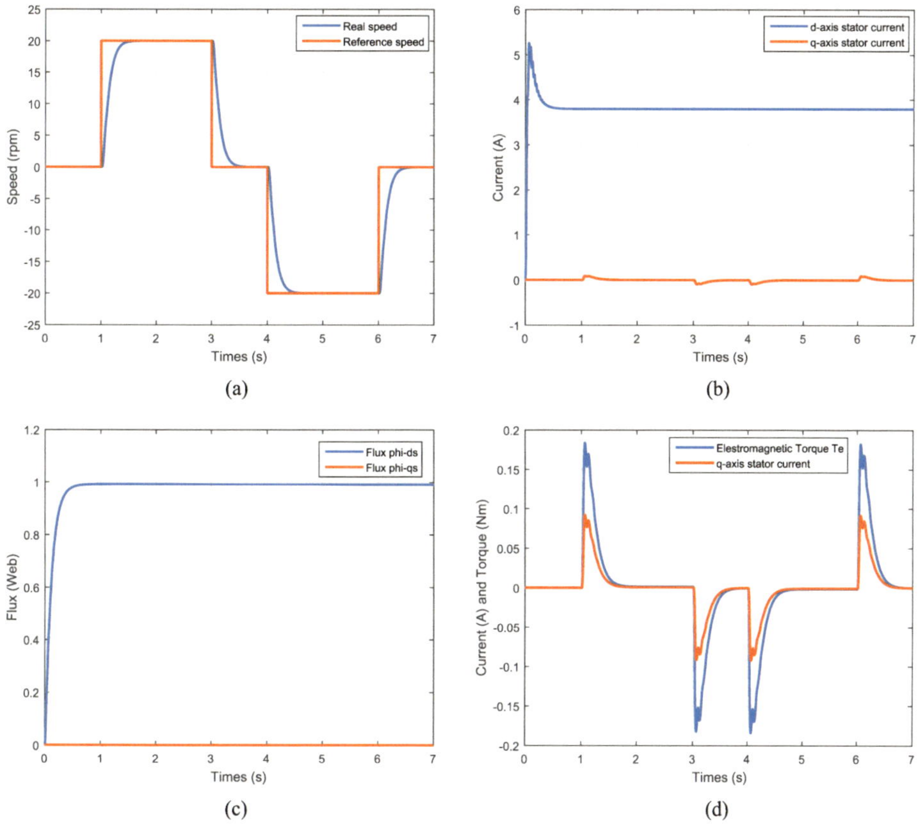

Figure 6.
Indirect stator flux-oriented controlled (ISFOC) induction motor drive for low speed.

The first is called the reference model since it does not introduce the quantity to be estimated and the second is the adjustable one. The error between these two models drives an adaptation mechanism that updates the parameters of the adjustable model to make them converge to the real parameters [11, 12]. **Figure 7** shows the basic block diagram of the MRAS method.

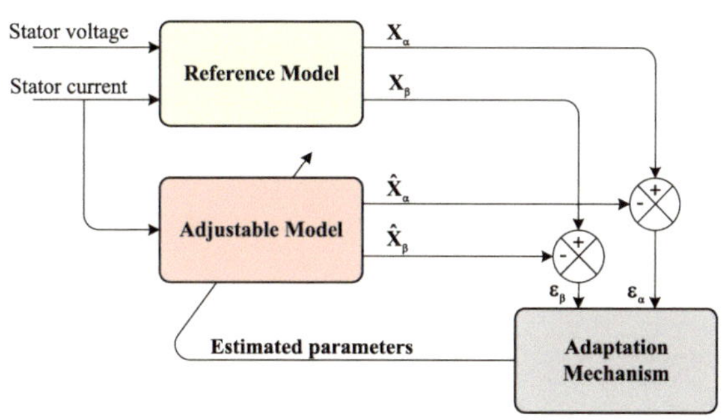

Figure 7.
The block diagram of the MRAS technique.

5.2 Simultaneous speed and rotor resistance estimation

For the estimation of the speed and the rotor resistance, it is judicious to use a stationary reference frame (α,β). This transformation does not use the position of the rotor, which is estimated by the MRAS strategy [11, 13].

The stator voltages can be described by the equation:

$$\begin{cases} v_{\alpha s} = R_s i_{\alpha s} + \dfrac{d\phi_{\alpha s}}{dt} \\ v_{\beta s} = R_s i_{\beta s} + \dfrac{d\phi_{\beta s}}{dt} \end{cases} \tag{27}$$

The rotor voltages can be described by the equation:

$$\begin{cases} 0 = R_r i_{\alpha r} + \dfrac{d\phi_{\alpha r}}{dt} + \omega\phi_{\beta r} \\ 0 = R_r i_{\beta r} + \dfrac{d\phi_{\beta r}}{dt} - \omega\phi_{\alpha r} \end{cases} \tag{28}$$

With

$$\begin{cases} \phi_{\alpha r} = L_r i_{\alpha r} + M i_{\alpha s} \\ \phi_{\beta r} = L_r i_{\beta r} + M i_{\beta s} \end{cases} \tag{29}$$

By using Eqs. (28) and (29) are rewritten under the following new form:

$$\begin{cases} 0 = -L_s D i_{\alpha s} - E i_{\beta s} + D\phi_{\alpha s} + \omega L_r \phi_{\beta s} \\ 0 = E i_{\alpha s} - L_s D i_{\beta s} - \omega L_r \phi_{\alpha s} + D\phi_{\beta s} \end{cases} \tag{30}$$

where $D = R_r + p\sigma L_r$. and $E = \sigma L_s L_r \omega$

According to Eqs. (28) and (29), is it possible to identify ω and R_r. Afterward, we look to illustrate the components of the stationary reference frame stator flux ($\phi_{\alpha s}$,$\phi_{\beta s}$) in terms of accessible stator currents ($i_{\alpha s}$,$i_{\beta s}$) and stator voltages ($v_{\alpha s}$,$v_{\beta s}$). Thus, two independent stator flux estimators are built. The integration of Eq. (27) gives the first estimator stator flux estimator, whereas Eq. (29) determines the second estimator. Knowing that Eq. (27) does not involve include the ω and R_r parameters. Therefore, it can be considered a reference model of the IM. However, Eq. (30) comprises the parameters ω and R_r, which is considered an adjustable model.

The suitable adaptation mechanism that generates the estimation of ω and R_r, for the adjustable model is driven by the error between the states of the two models given by Eq. (31). It is noticed well that the feedback block *W*, which is given by Eq. (32) is different for the speed estimation and rotor estimation, then so two adaptation mechanisms (1 and 2) comes up.

$$\begin{cases} \varepsilon_\alpha = \phi_{\alpha s} - \hat{\phi}_{\alpha s} \\ \varepsilon_\beta = \phi_{\beta s} - \hat{\phi}_{\beta s} \end{cases} \tag{31}$$

Eq. (6) can be presented as following [11, 14]:

$$p[\varepsilon] = [A][\varepsilon] + [W] \tag{32}$$

where W is the feedback block, which establishes the input of the linear block.

- For the speed estimation.

$$A = \begin{bmatrix} -\frac{1}{\tau_r} & -\omega \\ \omega & -\frac{1}{\tau_r} \end{bmatrix} \text{ and } W = W_1 = \begin{bmatrix} -\hat{\phi}_{\beta s} + \sigma L_s i_{\beta s} \\ \hat{\phi}_{\alpha s} - \sigma L_s i_{\alpha s} \end{bmatrix} (\omega - \hat{\omega})$$

Eqs. (31) and (32) constitute a nonlinear feedback system represented by **Figure 8**. Indeed, this system can be schematized by a linear block described by the transfer matrix $H(p) = (p[I] - [A])^{-1}$ and a nonlinear part of the input ε(t) and the output W(ε,t).

- For the rotor resistance estimation.

$$A = \begin{bmatrix} -\frac{R_r}{L_r} & -\omega \\ \omega & -\frac{R_r}{L_r} \end{bmatrix} \text{ and } W = W_2 = \begin{bmatrix} \frac{-\hat{\phi}_{\alpha s} + L_s i_{\alpha s}}{L_r} \\ \frac{-\hat{\phi}_{\beta s} + L_s i_{\beta s}}{L_r} \end{bmatrix} (R_r - \hat{R}_r)$$

Eqs. (31) and (32) constitute a nonlinear feedback system represented by **Figure 9**. Indeed, this system can be schematized by a linear block described by the transfer matrix $H(p) = (p[I] - [A])^{-1}$ and a nonlinear part of the input ε(t) and the output W(ε,t).

The MRAS strategy consists to use the adaptation mechanism 1 to estimate ω. Subsequently, adaptation mechanism 2 is used to estimate R_r (**Figure 10**).

The asymptotic operation of the adaptation mechanism is fulfilled by a simplified condition, which is $[\varepsilon(\infty)]^T = 0$. To be regarded as hyper-stable, the system of negative feedback must satisfy the inequality of Popov [15]:

$$\int_0^{t_1} [\varepsilon]^T [W] dt \geq -\gamma^2 \text{ for all } t_1 \geq 0 \tag{33}$$

With γ is a negative constant.

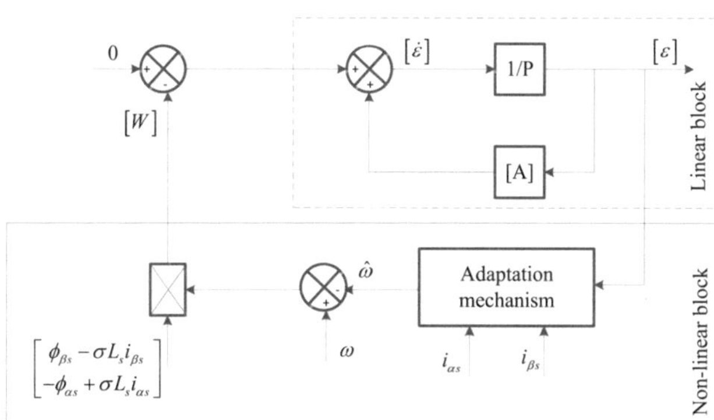

Figure 8.
Nonlinear feedback system for speed estimation.

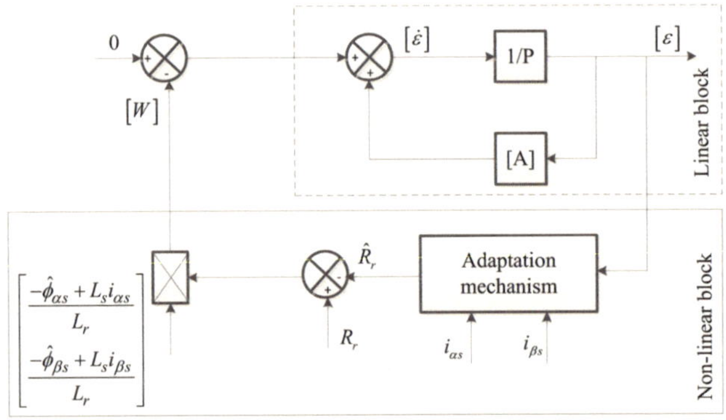

Figure 9.
Nonlinear feedback system for rotor resistance estimation.

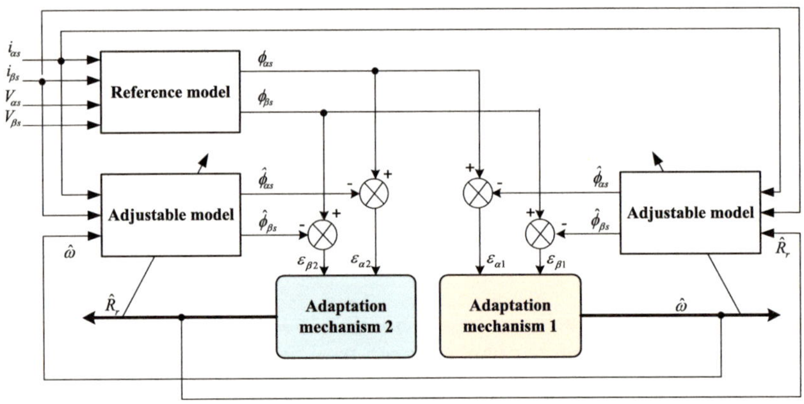

Figure 10.
Structure of the simultaneous speed and rotor resistance estimation by MRAS technique.

5.3 Speed adaptation mechanism

Adaptation mechanism 1 demonstrates that the estimated speed $\hat{\omega}$ is a function of the error [ε]. Then, it can be shown in the following form:

$$\hat{\omega} = \frac{1}{p}A_1 + A_2 = \int_0^{t_0} A_1([\varepsilon_1])dt + A_2([\varepsilon_1]) \tag{34}$$

where A_1 and A_2 are a nonlinear function of $\varepsilon_{\alpha 1}$, $\varepsilon_{\beta 1}$.
By using the expression of W, Eq. (33) is equivalent to

$$\int_0^{t_0} \left\{ \left[\varepsilon_{\alpha 1}\left(-\hat{\phi}_{\beta s} + \sigma L_s i_{\beta s}\right) \right] + \left[\varepsilon_{\beta 1}\left(\hat{\phi}_{\alpha s} - \sigma L_s i_{\alpha s}\right) \right] \right\} (\omega - \hat{\omega}) dt \geq -\gamma^2 \tag{35}$$

$$\text{With}: \begin{cases} A_1 = K_2\left[\left(\phi_{\beta s}\hat{\phi}_{\alpha s} - \phi_{\alpha s}\hat{\phi}_{\beta s} - \left(i_{\alpha s}\varepsilon_{\beta 1} - i_{\beta s}\varepsilon_{\alpha 1}\right)\sigma L_s \right] \\ A_2 = K_1\left[\left(\phi_{\beta s}\hat{\phi}_{\alpha s} - \phi_{\alpha s}\hat{\phi}_{\beta s} - \left(i_{\alpha s}\varepsilon_{\beta 1} - i_{\beta s}\varepsilon_{\alpha 1}\right)\sigma L_s \right] \end{cases}$$

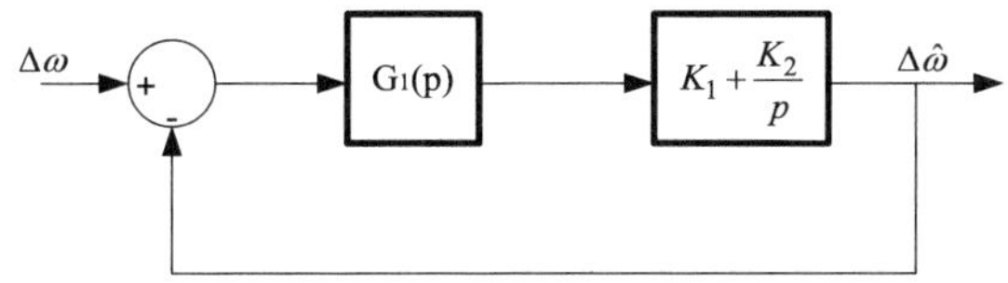

Figure 11.
Synthesis of the speed corrector.

where K_1 and K_2 are called adaptation gains, which are constants and positive.

Eq. (34) is built around a PI regulator. **Figure 11** represents the result of the synthesis of the speed regulator [11, 14].

To study the dynamic response of the estimated speed by the MRAS method, it is necessary to linearize the stator and the rotor equations around an operating point. Thus, the error variation ε is expressed as follows [11, 14]:

$$\Delta\varepsilon = \frac{\phi_{\alpha s0}\Delta\phi_{\alpha s0} + \phi_{\beta s0}\Delta\phi_{\beta s0}}{\sqrt{\phi_{\alpha s0}^2 + \phi_{\beta s0}^2}} - \frac{\hat{\phi}_{\alpha s0}\Delta\hat{\phi}_{\alpha s0} + \hat{\phi}_{\beta s0}\Delta\hat{\phi}_{\beta s0}}{\sqrt{\hat{\phi}_{\alpha s0}^2 + \hat{\phi}_{\beta s0}^2}} \tag{36}$$

The transfer function representing the variation ratio Δε compared with $\Delta\hat{\omega}$ is written as:

$$\left.\frac{\Delta\varepsilon}{\Delta\hat{\omega}}\right|_{\Delta\omega \to 0} = \frac{(p+\beta_r)\phi_0^2}{\left((p+\beta_r)^2 + \omega^2\right)} \tag{37}$$

In a steady operation, one will have: $\phi_0^2 = \phi_{\alpha s0}^2 + \phi_{\beta s0}^2$ and $\omega = \hat{\omega}$.

Then, by neglecting the effect of the slip pulses ω_{sl}, the transfer function of $G_1(p)$ will be:

$$G_1(p) = \frac{\phi_0^2}{p + \frac{1}{\tau_r}} \tag{38}$$

Hence, the transfer function of the direct chains is written:

$$F(p) = \frac{\phi_0^2(K_1p + K_2)}{\left(p + \frac{1}{\tau_r}\right)p} \tag{39}$$

The transfer function F(p) of the speed estimator is influenced by PI regulator coefficients K_1 and K_2, real speed, and induction motor parameters. To highlight the influence of the real speed on the stability of the estimator, the place of the zeros and the poles for the various values of ω are presented. When the rotor speed varies in the interval from 0 to 314 rad/s, the MRAS estimator poles location is presented in **Figure 12**. The results demonstrate that the variation speed does not affect the stability of the system.

5.4 Rotor resistance adaptation mechanism

Adaptation mechanism 2 demonstrates that the estimated rotor resistance is a function of the error [ε]. Then, it can be shown in the following form [11, 14]:

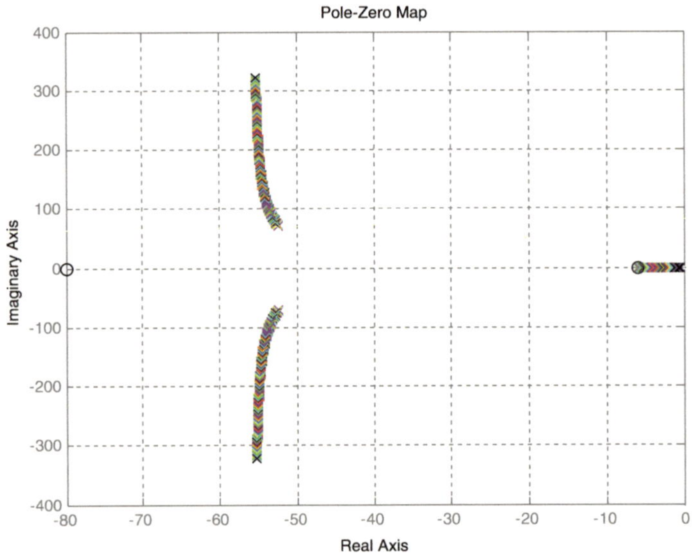

Figure 12.
Evolution of the MRAS estimator poles and zeros according to the speed.

$$\hat{R}_r = \frac{1}{p}A_3 + A_4 = \int_0^{t_0} A_3([\varepsilon_2])dt + A_4([\varepsilon_2]) \tag{40}$$

where A_3 and A_4 are a nonlinear function of $\varepsilon_{\alpha 2}$, $\varepsilon_{\beta 2}$.
Using the expression of W_2, Eq. (33) is equivalent to:

$$\int_0^{t_0} \left\{ \left[\varepsilon_{\alpha 2} \left(\frac{-\hat{\phi}_{\alpha s} + L_s i_{\alpha s}}{L_r} \right) \right] + \left[\varepsilon_{\beta 2} \left(\frac{-\hat{\phi}_{\beta s} + L_s i_{\beta s}}{L_r} \right) \right] \right\} (R_r - \hat{R}_r) dt \geq -\gamma^2 \tag{41}$$

$$\text{with :} \begin{cases} A_3 = K_4 \left[\left(\frac{-\hat{\phi}_{\alpha s} + L_s i_{\alpha s}}{L_r} \right) \varepsilon_{\alpha 2} + \left(\frac{-\hat{\phi}_{\beta s} + L_s i_{\beta s}}{L_r} \right) \varepsilon_{\beta 2} \right] \\ A_4 = K_3 \left[\left(\frac{-\hat{\phi}_{\alpha s} + L_s i_{\alpha s}}{L_r} \right) \varepsilon_{\alpha 2} + \left(\frac{-\hat{\phi}_{\beta s} + L_s i_{\beta s}}{L_r} \right) \varepsilon_{\beta 2} \right] \end{cases}$$

where K_3 and K_4 are called adaptation gains, which are constant and positive.
Eq.(42) presents a transfer function that connects $\Delta\varepsilon$ with $\Delta\hat{R}_r$:

$$\left. \frac{\Delta \varepsilon}{\Delta \hat{R}_r} \right|_{\Delta R_r \to 0} = \frac{(p + \delta_0)\left(|\phi_0|^2 - L_s(\phi_{\alpha s0} i_{\alpha s0} + \phi_{\beta s0} i_{\beta s0}) \right)}{L_r\left((p + \delta_0)^2 + \omega^2 \right) |\phi_0|} \tag{42}$$

In a steady operation, one will have: $\phi_0^2 = \phi_{\alpha s0}^2 + \phi_{\beta s0}^2$ and $R_r = \hat{R}_r$

The functional diagram in the closed loop of the rotor resistance estimation by the MRAS method is given in **Figure 13**.

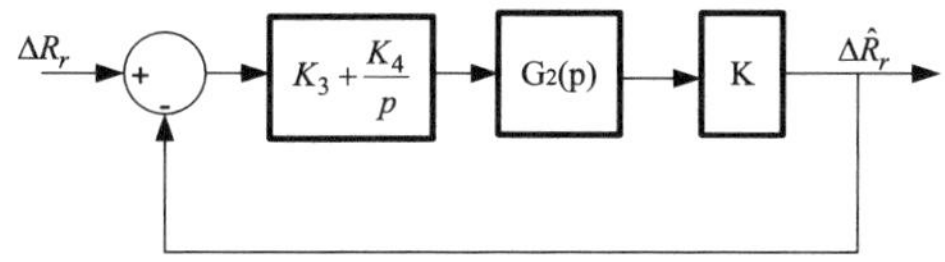

Figure 13.
Functional diagram of the rotor resistance estimation.

In this functional diagram, the expression of K is given by [14]:

$$K = \frac{|\phi_0|^2 - L_s(\phi_{\alpha s0} i_{\alpha s0} + \phi_{\beta s0} i_{\beta s0})}{L_r |\phi_0|} \tag{43}$$

$$G_2(p) = \frac{p + \frac{R_r}{L_r}}{\left(p + \frac{R_r}{L_r}\right)^2 + \omega^2} \tag{44}$$

Two complex poles (p_1 and p_2) are presented in the transfer function $G_2(p)$:

$$p_1 = -\frac{R_r}{L_r} + j\omega \tag{45}$$

and

$$p_2 = -\frac{R_r}{L_r} - j\omega \tag{46}$$

We note that the stability of the function is confirmed because the real part of the two complex poles p_1 and p_2 are negative. Furthermore, the application of a PI regulator allowed us to have a fast convergence and a null static error in the steady-state operation.

Figure 14 shows the layout of the root locations related to the transfer function $G_2(p)$ for 157 rad/s. It characterizes the dynamics of the control system. However, for this function, there are two poles and a zero, which are all located in the left half-plan of the complex plan. Moreover, the poles p_1 and p_2 are located at $-5.94 \pm 157j$, which characterize the dynamics of the system. Then, to guarantee a good tracking performance and fast transient response, the zero placement of the PI regulator must be located on the real axis at -5.94. Thus, K_3 must be taken of a rather large value and the ratio K_4/K_3 must be equal to the negative real part of the complex poles of the systems [16, 17] (K_3 = 500 and K_4/K_3 = 5,94). This allows us to have an error in the rotor resistance that decreases the fast exponential form with time without oscillation.

Figure 15 shows the zero location of the controller PI (K_4/K_3) can be selected for the value of the rotor resistance. Otherwise, this figure presents the location of the poles of $G_2(p)$ with various values of R_r.

5.5 Stator flux estimation

According to system Eq. (2), the two flux $\phi_{\alpha s}$ and $\phi_{\beta s}$ are determined through the integration of the back EMF of the machine. The estimated flux is obtained by a pure

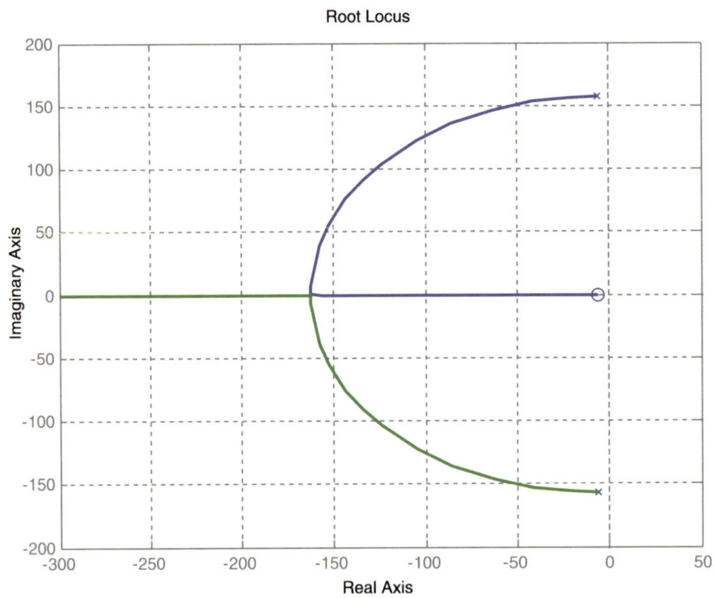

Figure 14.
Poles and zeros locations of G2 (p).

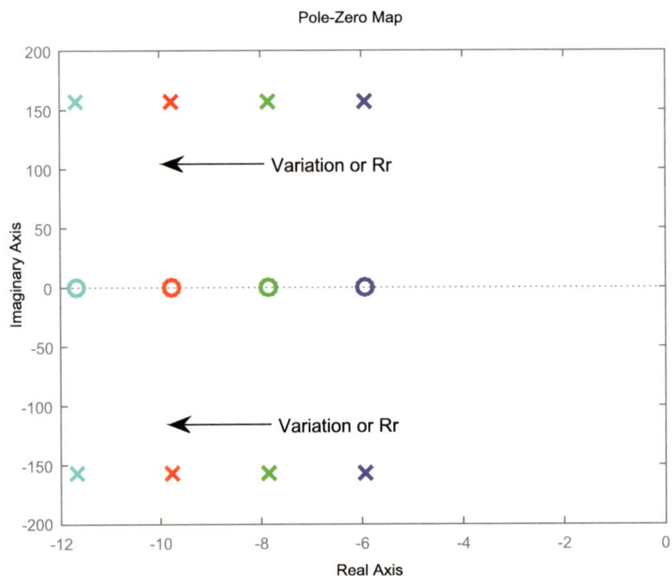

Figure 15.
Poles and zeros locations of G2 (p) for different values of Rr.

integration and can cause significant problems, in particular, at the voltage shift (DC offset), at the uncertain parameters, and at low frequency due to the noise sensor. Thus, to overcome these problems, the pure integrator is changed with a low-pass filter (LPF) [18]. In this chapter, we propose to replace the pure integrator with a programmable cascaded low pass filter (PCLPF). This approach has already been proposed to control the IM without a mechanical sensor [19]. Indeed, a PCLPF includes a few blocks of these filters. Each block decreases the effect of the voltage shift (DC offset). In one PCLPF, their gains as well as their cut-off frequencies are given according to the stator frequency.

Therefore, two methods are proposed. A Modified Programmable Cascaded Low-Pass Filters (MPCLPF) present the first method, the estimate precision is improved by eliminating the gain in the calculation phase. The second method is called the Extended PCLPF (EPCLPF) [20], the estimator behavior is modified by the elimination of the gain G in the calculation phase and the extension low-pass filters stages (LPF stages). This method can be applied only to the area of a very low speed.

5.5.1 MPCLPF filter

The phase angle θ_s is calculated by dividing $\phi_{\beta s}(p)$ by $\phi_{\alpha s}(p)$. Thus, G is a common factor in the two equations ($\phi_{\alpha s}(p)$ and $\phi_{\beta s}(p)$) and can be eliminated. (by simplification). However, it can be announced that:

$H'(p)$, which is defined as the transfer function of MPCLPF is presented by Eq. (47).

$$H'(p) = \left(\frac{1}{\tau \cdot p + 1}\right)^3 \tag{47}$$

$$F_\alpha(p) = (v_{\alpha s}(p) - R_s i_{\alpha s}(p))H'(p) \tag{48}$$

$$F_\beta(p) = (v_{\beta s}(p) - R_s i_{\beta s}(p))H'(p) \tag{49}$$

$$\theta_s = \tan^{-1}\left(\frac{F_\beta(t)}{F_\alpha(t)}\right) = \tan^{-1}\left(\frac{\ell^{-1}(F_\beta(p))}{\ell^{-1}(F_\alpha(p))}\right) \tag{50}$$

5.5.2 Extended programmable cascaded low pass filter (EPCLPF)

As previously mentioned, the reduction of the continuous voltage shift (DC offset) is an effective means for sensorless control in the low-speed region [20]. As the output depends on the gain of the PCLPF, this list should be reduced. Here, it is necessary to take the relation between the amplitude of the gain and the order of the PCLPF into account. For a number of N LPF stages presented in **Figure 16**, the time-constant of the filters, the calculations of the gain and the transfer function are given below:

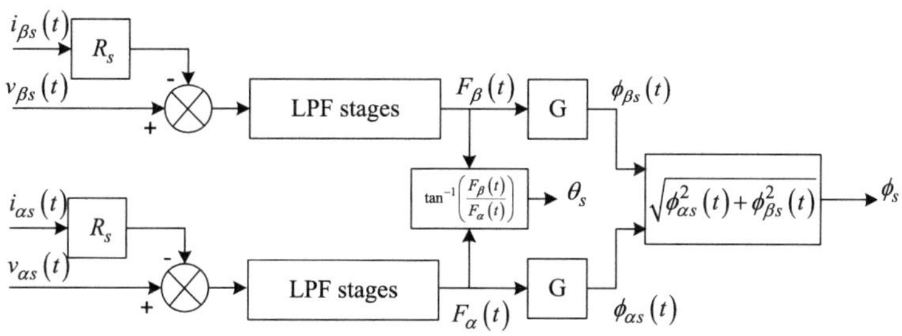

Figure 16.
Diagram of N floors of the programmable-low-flow filter.

$$\tau = \frac{1}{\omega_e} \tan\left(\frac{\pi}{2 \cdot n}\right) \tag{51}$$

$$G = \frac{1}{\omega_e} \sqrt{\left[1 + (\omega_e \tau)^2\right]^n} \tag{52}$$

By using Eqs. (51) and (52) becomes:

$$G = \frac{1}{\omega_e} \sqrt{\left[1 + \tan^2\left(\frac{\pi}{2 \cdot n}\right)\right]^n} \tag{53}$$

$$H(p) = G\left(\frac{1}{\tau \cdot p + 1}\right)^n \tag{54}$$

5.6 Simulation results

The simultaneous estimation algorithm by MRAS method has been simulated using Matlab-Simulink Software. **Figure 17** shows the synoptic of simultaneous estimation in sensorless ISFOC IM drive based on MRAS scheme. However, the bloc diagram is made up of a PWM voltage source inverter, a coordinate translator, and a speed controller. A field orientation mechanism and an induction motor. Moreover, the gains of the PI current controller and IP gains speed controller are calculated and tuned at each sampling time, according to the new estimated rotor speed and rotor resistance.

For the simulation of the proposed algorithm, we used a 3 kW induction motor knowing that these parameters are given in the Appendix. Furthermore, the sampling time for control algorithms computation and simultaneous estimation is

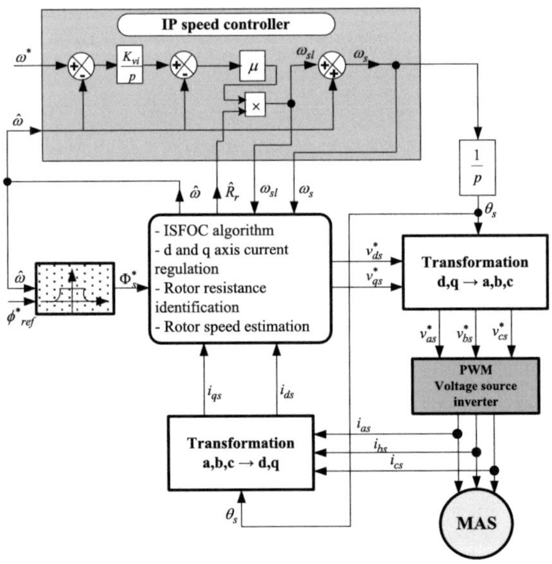

Figure 17.
Synoptic of ISFOC-oriented induction motor drive system with simultaneous estimation.

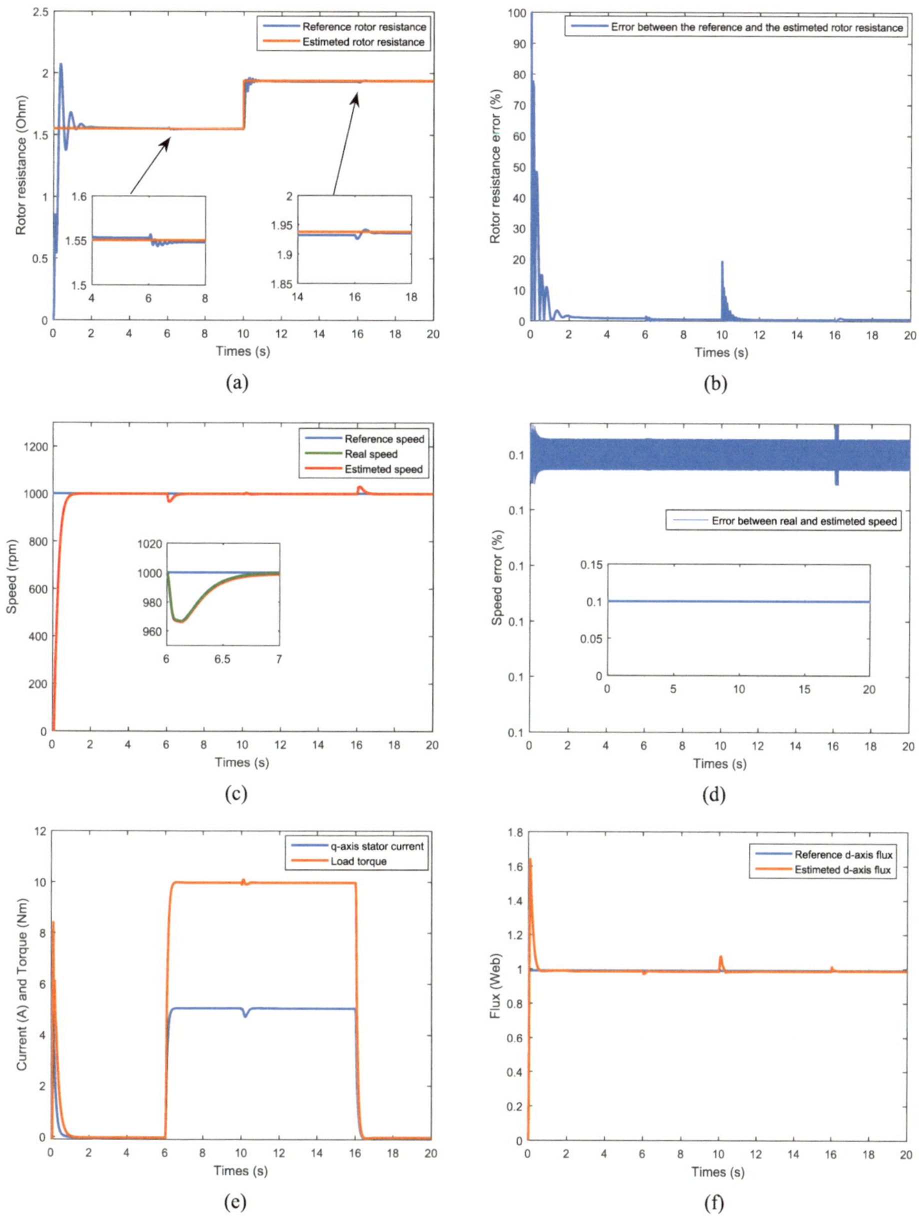

Figure 18.
Identification performance at nominal speed and load torque: (a) estimated and reference rotor resistance, (b) rotor resistance error, (c) reference speed, real speed, and estimated speed (d) estimated and reference flux, (e) iqs stator current and load torque, and (f) φds estimated and reference stator flux.

used as 0.2 ms. Moreover, this sensorless ISFOC drive simulation was performed under various load and speed conditions. Thus, in **Figure 18**, at a speed of 1000 rpm and with a load torque, R_r was chosen to be equal to 1.55 Ω and increased by 50% from its initial value at t = 10 s. Knowing that the torque is applied with 10 Nm at t = 6 s and canceled at time t = 16 s. The two trajectories of estimated and reference rotor resistance coincide fairly well and a very good coincidence is reached. We also note that the error between the real and estimated rotor resistance is given in the **Figure 18(b)**.

The increase of the rotor resistance agrees with an eventual heating of the rotor winding. The obtained result demonstrates that even if the rotor resistance changes, the proposed ISFOC procedure still gives a good estimate of this parameter.
Figure 18(c) shows the simulation result of the reference, the estimated and the real speed using the IP controller. When the speed control changes from zero to 1000 rpm and the load torque applied and removed at t = 6 and 16 s, respectively, the error between the real and estimated rotor speed presented in **Figure 18(d)** is less than 0.15%. Furthermore, **Figure 18(e)** shows the measurement of the load torque and the q-axis stator current. The electromagnetic torque is estimated by the use of a stator current measurement. Moreover, the convergence of the estimated size of ϕ_{ds} stator flux to its reference is proved by the **Figure 18(f)**.

To show the robustness of the indirect control by the orientation of the stator flux, we applied a low-speed reference with the inversion of the rotation direction (0tr/mn → 150 pm → −15 rpm).

However, the estimated rotor resistance, as shown in **Figure 19(a)**, matches the actual rotor resistance of the machine with a low error (**Figure 19(b)**) and so we obtain a robust control performance. Furthermore, in **Figure 19(a)**, the three trajectories of estimated, real and reference rotor speed coincide fairly well and a good coincidence is reached. The results show that the ISFOC has a good tracking performance even at low speeds.

6. Experimental results

A prototype of the sensorless control with estimation of the rotor resistance of an asynchronous machine has been produced. As a result, experimental tests were carried out to validate the proposed algorithm and demonstrated the effectiveness and robustness of the MRAS technique. The experimental setup is composed of squirrel-cage IM a 3 kW, a voltage source inverter (VSI), pulse width modulation (PWM) signals to control the power modules generated by dSpace system, a hall-type sensors stator currents and stator voltages and a load generated through a magnetic power brake (**Figure 20**). Knowing that the experimentation has been achieved by using Matlab-Simulink and dSpace DS1104 real-time controller board.

Two cases will be presented for experimental validation:

- First case: A speed reference of 15 rpm, and load torque (20 Nm) applied at t = 7 s and canceled at time t = 13 s.

- Second case: A speed reference of 0 → 15 rpm → −15 rpm.

• The first case:

The experimental tests illustrated by **Figure 21** (**a**) show that the estimated speed and the actual speed converge to that of the reference. One may also notice low oscillations because of the reliability and fidelity of the sensor for the selected speed.

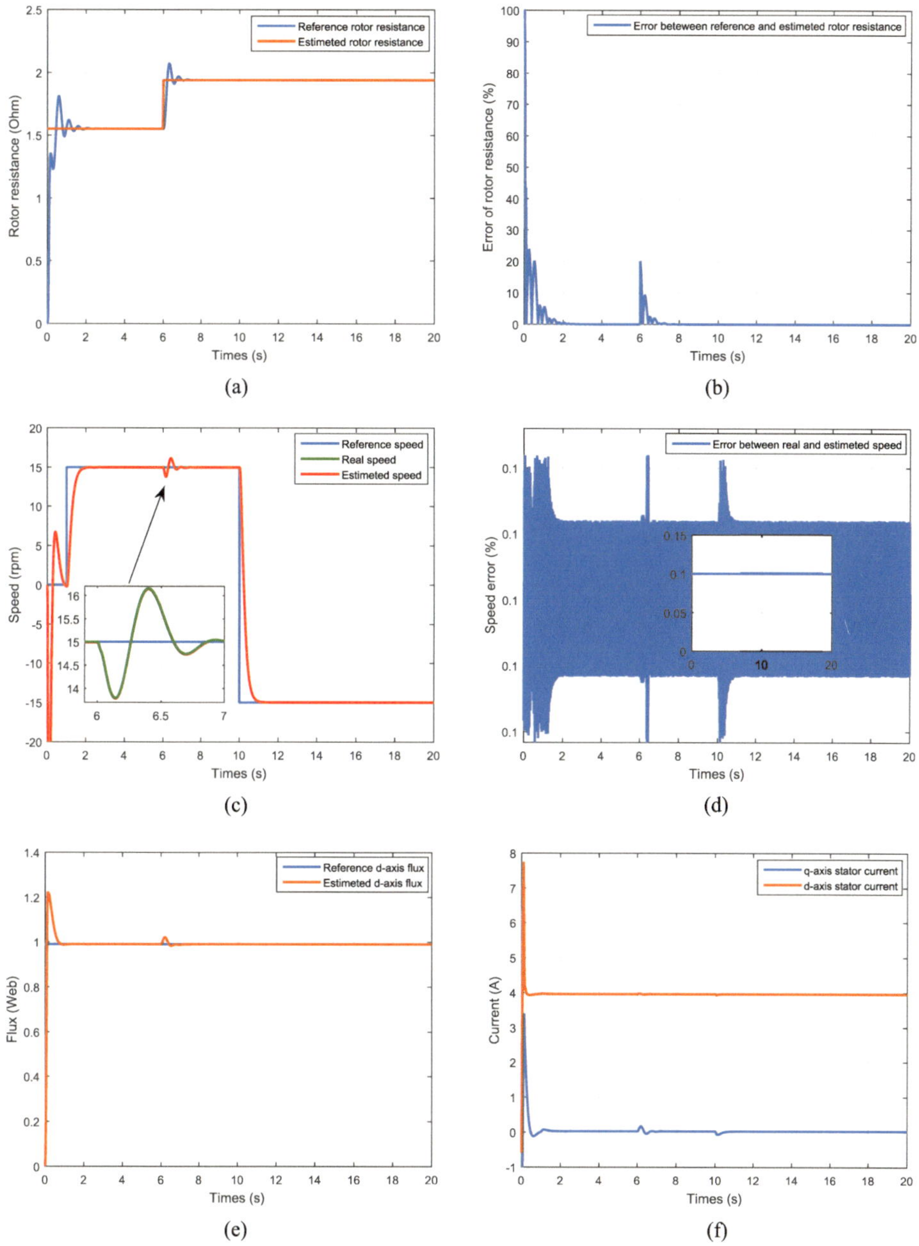

Figure 19.
Identification performance at nominal speed and load torque: (a) estimated and reference rotor resistance, (b) rotor resistance error, (c) reference speed, real speed, and estimated speed (d) estimated and reference flux, (e) iqs stator current and load torque, and (f) ϕds estimated and reference stator flux.

The sensor of the used velocity is an incremental coder with 1024 points. It actually presents a speed resolution: $n_{res} = \frac{60}{4 \cdot number\ point\ of\ codeur \cdot T_e}$ where Te is the sampling step; hence, this value is significant for low speed, and consequently, the error

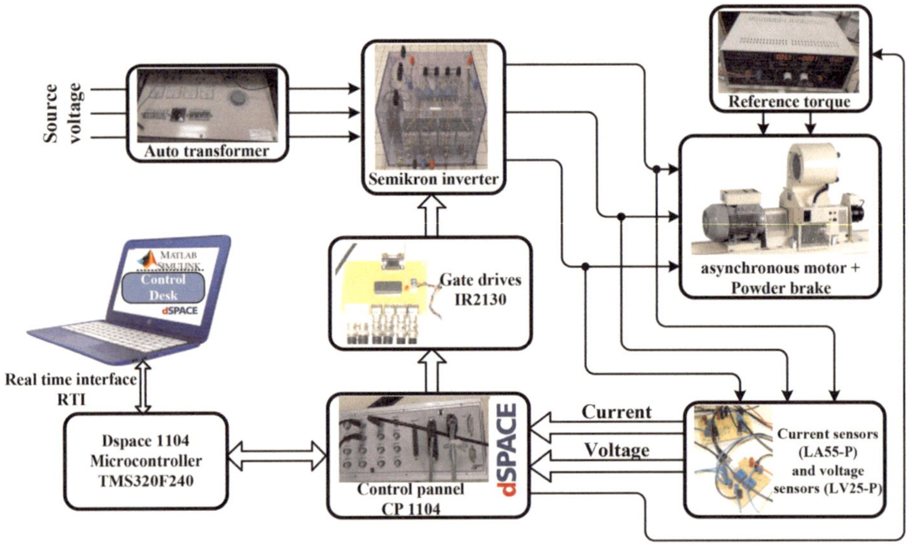

Figure 20.
Experimental setup.

increases. Furthermore, we use an incremental encoder with 1024 pulses per revolution as a speed sensor, and consequently, at a low-speed operation, the speed precision decreases. Thus, in **Figure 21(b)** the error between the estimated and the real speed does not exceed 2 rpm in the steady state. **Figure 21(c)** presents the convergence of the estimated rotor resistance R_r to its reference value. The error between the estimated and the reference rotor resistance is given in **Figure 21(d)**. Furthermore, the proportionality between the q-axis stator current and the electromagnetic torque is confirmed in **Figure 21(e)**. Once again, **Figure 21(g)** proves that the flux $\phi_{\alpha s}$ and $\phi_{\beta s}$ are in a quadratic phase. The stator currents of the d-axis and q-axis are shown in **Figure 21(f)**. Finally, **Figure 21(h)** proves the concordance of the stator currents i_{as}, i_{bs}, and i_{cs}.

• The second case:

In **Figure 22**, the sensorless ISFOC induction motor tests are carried out to show the effectiveness and robustness of the proposed scheme and the behaviors of the motor during speed transients. In **Figure 22(a)**, the speed reversal test is performed and we note the convergence of the real and estimated speed toward its reference value. Therefore, **Figure 22(b)** shows that the error between the estimated and the real speed does not exceed 2 rpm in the steady state. **Figure 22(c)** shows that the convergence of the estimated value of rotor resistance R_r to its reference is proved. This convergence is confirmed by **Figure 22(d)**, which represents the error between the estimated and the reference value. Moreover, **Figure 22(e)** proves that the flux $\phi_{\alpha s}$ and $\phi_{\beta s}$ are in a quadratic phase. The convergence of the estimated stator current i_{qs} to its reference is confirmed by **Figure 22(f)**, the same thing for **Figure 22(g)**, which presents the i_{ds} stator current.

Finally, the **Figure 22(h)** proves the concordance of the stator currents i_{as}, i_{bs}, and i_{cs}.

These results confirm the effectiveness, robustness, and good tracking performance of the ISFOC based on the MRAS scheme.

• The first case:

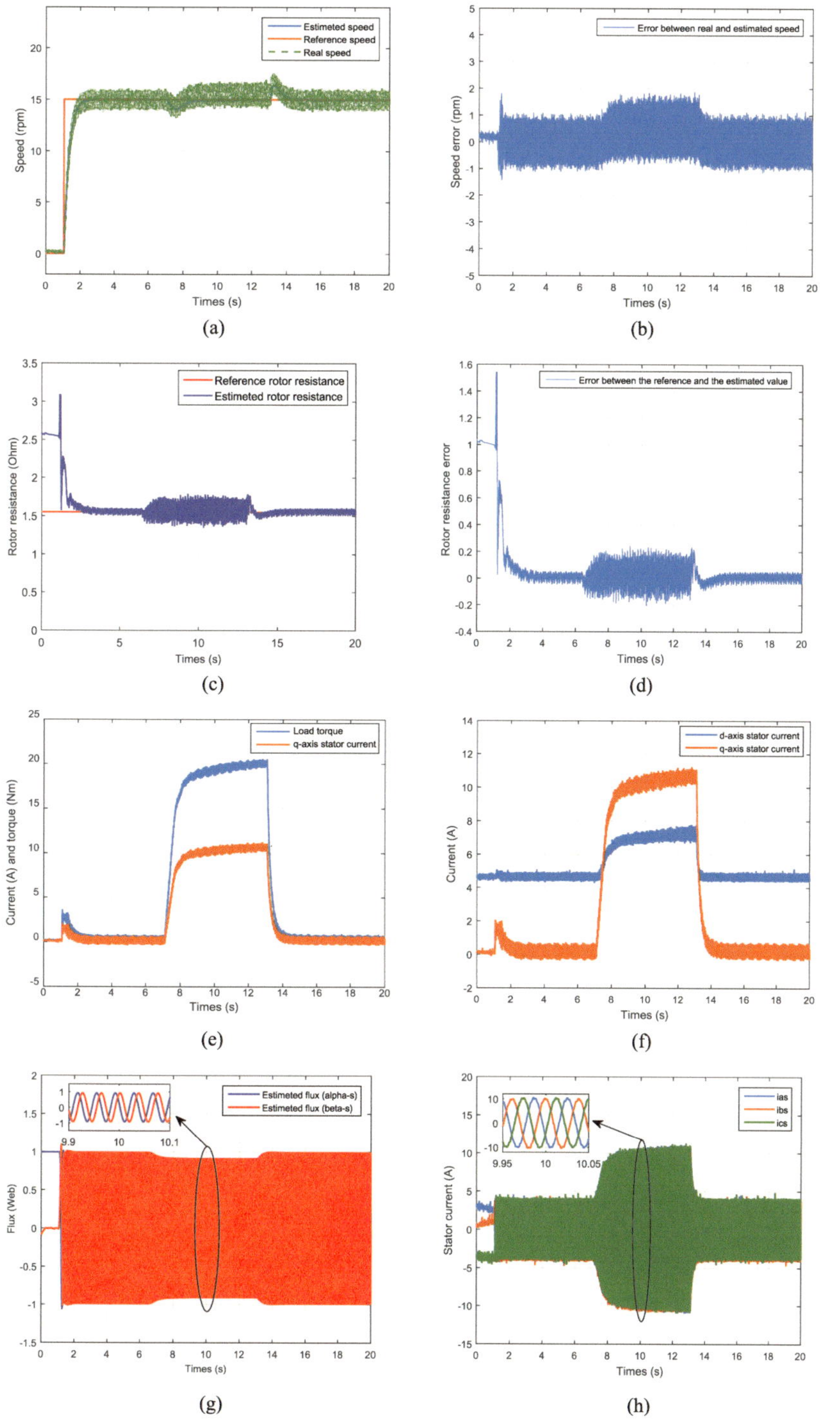

Figure 21.
Experimental results of step response (15 rpm) with load torque applied and removed at t = 7 s and 13 s, respectively.

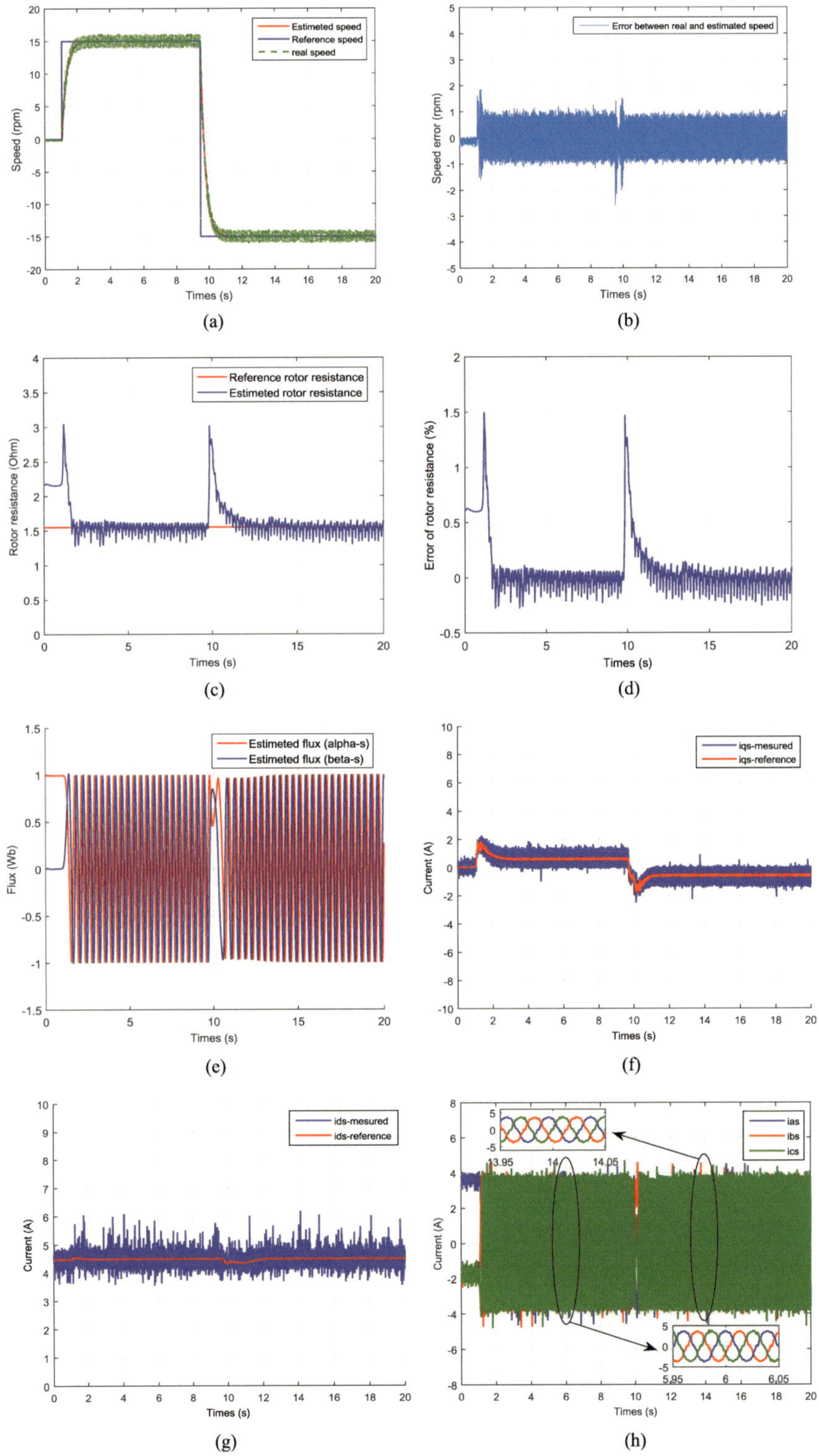

Figure 22.
Experimental results of step response ω = 15 rpm with inverting direction at the moment t = 9.5 s.

7. Conclusion

A simultaneous estimation of speed and rotor resistance in sensorless ISFOC induction motor drive based on the MRAS scheme has been presented. The variation of the rotor resistance is taken into account in order to guarantee the stability of the control system in all speed ranges with high-performance sensorless speed. The simultaneous estimation by the MRAS method is ensured by measuring the stator currents and voltages. Furthermore, the PI current and IP speed controller gains are online performed at each sampling time with the estimated value of the rotor resistance in order to provide optimal performance with guaranteed stability of the MRAS scheme.

From the simulation results, the simultaneous speed and rotor resistance estimation has proved the effectiveness and robustness of the sensorless ISFOC induction motor drive.

We have validated the online estimation for the rotor speed and the rotor resistance of an induction motor operating in an indirect stator field-oriented control system based on the MRAS method. The experimental results prove the robustness of this simultaneous estimation and the compliance with simulation results. The analyzed closed-loop stability of the proposed technique has also been proved through the Lyapunov stability theory. The experimental results show that the rotor resistance is sensitive to load variation. More importantly, the validity of the proposed sensorless ISFOC of the induction motor drive was proven by experiments for a wide range of speed and for a very low speed. More importantly, all experimental results confirm the good dynamic performances of the developed drive systems and show the validity of the suggested method. The obtained results show that the rotor resistance R_r is sensitive to load variation.

A. Appendix

Nomenclature

v_{ds}, v_{qs}, i_{ds}, i_{qs}	d, q-axis stator voltage and current components
ϕ_{ds}, ϕ_{qs}	d, q-axis stator flux components
$\phi_{\alpha s}$, $\phi_{\beta s}$	α, β-axis stator flux components
R_r, R_s	rotor and stator resistance
L_r, L_s	rotor and stator self-inductance
M	mutual inductance
n_p	number of pole pairs
ω_s, ω	synchronous and rotor angular speed
ω_{sl}	slip angular speed (ω_s-ω)
T_e, T_l	electromagnetic and load torque
J	moment of inertia
f	friction constant
τ_s, τ_r	stator and rotor time constant
k_{vi}, k_{vp}	integral an proportional gain of the IP speed controller
k_{ip}, k_{ip}	integral and proportional gain of the PI current controller
σ	total leakage constant
^, *	estimated and reference value

$p = \frac{d}{dt}$ differential operator

List of motor specification and parameters: 380 V, 3 KW, 4 poles, 1430 rpm, R_s = 2.3 Ω; R_r = 1.55 Ω; L_s = L_r = 0.261 H; M = 0.245 H; f = 0.002 Nm s/rd; J = 0.03 kg m^2.

Author details

Youssef Agrebi Zorgani[1,2]*, Soufien Hajji[1], Yessine Koubaa[1] and Mohamed Boussak[3]

1 Laboratory of Sciences and Techniques of Automatic Control and Computer Engineering (Lab-STA), National School of Engineering of Sfax, University of Sfax, Sfax, Tunisia

2 Higher Institute of Technological Studies of Sfax, Sfax, Tunisia

3 Laboratoire d'Informatique et Systèmes (LIS), Ecole Centrale de Marseille (ECM), Rechnopôle de Chateau Gombert, Marseille, France

*Address all correspondence to: agrebi69@yahoo.fr

References

[1] Jemli M, Boussak M, Goussa M, Kamoun MBA. Fail-safe implementation of indirect filed oriented controlled induction motor drive. Simulation Practice and Theory Journal. 2000;**8**: 233-252

[2] Boldea I, Nasar SA. The Induction Machine Handbook. 1st ed. Boca Raton: CRC Press; 29 November 2001. p. 968. DOI: 10.1201/9781420042658

[3] Xing Y. A novel rotor resistance identification method for an indirect rotor flux-oriented controlled induction machine system. IEEE Transactions on Power Electronics. 2002;**17**:353-364

[4] Maiti S, Chakraborty C, Hori Y, Ta MC. Model reference adaptive controller based rotor resistance and speed estimation technique for vector controlled induction motor drive utilizing reactive power. IEEE Transactions on Industrial Electronics. 2008;**55**(2):594-601

[5] Akatsu K, Kawamura A. Online rotor resistance estimation using the transient state under the speed sensorless control of induction motor. IEEE Transactions on Power Electronics. 2000;**15**(3): 553-560

[6] Miyashita I, Fujita H, Ohmori Y. Speed sensorless instantaneous vector control with identification of secondary resistance. In: Proc. IEEE IAS Annual Meeting. 1991. pp. E130-E135

[7] Jiang J, Holtz J. High dynamic speed sensorless ac drive with on-line model parameter tuning for steady-state accuracy. IEEE Transactions on Industrial Electronics. 1997;**44**:240-246

[8] Kubota H, Matsuse K. Speed sensorless field-oriented control of induction motor with rotor resistance adaptation. IEEE Transactions on Industry Applications. 1994;**30**:1219-1224

[9] Abdessamed R, Kadjoudi M. Modélisation des machines électriques. Presse de l'université de Batna Algerie. 1997

[10] Barret F. Régimes Transitoires des Machines Tournantes Électrique. Edition Eyrolls Paris: Collection des études et recherches d'électricité de France; 1982

[11] Agrebi Y, Triki MK, Y and Boussak, M. Rotor speed estimation for indirect stator flux oriented induction motor drive based on MRAS scheme. Journal of Electrical Systems. 2007;**3**(3):131-114

[12] Agrebi-Zorgani Y, Koubaa Y, Boussak M. MRAS state estimator for speed sensorless ISFOC induction motor drives with Luenberger load torque estimation. ISA Transactions. 2016;**61**: 308-317

[13] Karlovský P, Linhart R, Lettl J. Sensorless determination of induction motor drive speed using MRAS method. In: Proceedings of the 8th International Conference on Electronics, Computers and Artificial Intelligence (ECAI). Ploiesti, Romania: IEEE Xplore; 2016. pp. 1-4

[14] Agrebi-Zorgani Y, Koubaa Y, Boussak M. Simultaneous estimation of speed and rotor resistance in sensorless ISFOC induction motor drive based on MRAS scheme. In: The XIX International Conference on Electrical Machines - ICEM 2010. Rome, Italy: IEEE; 2010. pp. 1-6

[15] Schauder C. Adaptive speed identification for vector control of induction motors without rotational

transducers. IEEE Transactions on Industrial Applications. 1992;**28**(5): 1054-1061

[16] Zaky MS. A stable adaptive flux observer for a very low speed sensorless induction motor drives insensitive to stator resistance variations. Ain Shams Engineering Journal, Production and Hosting by Elsevier. 2011;**2**(1):11-20

[17] Zaky MS. Stability analysis of speed and stator resistance estimators for sensorless induction motor drives. IEEE Transactions on Industrial Electronics. 2012;**59**(2):858-870

[18] Ghaderi A, Hanamoto T. Wide-speed-range sensorless vector control of synchronous reluctance motors based on extended programmable cascaded low-pass filters. IEEE Transactions on Industrial Electronics. 2011;**58**(6): 2322-2333

[19] Ghaderi A, Hanamoto T, Teruo T. A novel sensorless low speed vector control for synchronous reluctance motors using a block pulse function-based parameter identification. Journal of power. Electronics. 2006;**6**(3):235-244

[20] Ghaderi A, Hanamoto T, Tsuji T. Very low speed sensorless vector control of synchronous reluctance motors with a novel startup scheme. In: IEEE Applied Power Electronics Conference and Exposition APEC. Anaheim, CA, USA. 25 Feburary-1 March 2007

Chapter 9

Modeling and Simulation of a PMSG-Based Marine Current Turbine System with Inter-Turn Faults

Sana Toumi, Mohamed Benbouzid and Mohamed Faouzi Mimouni

Abstract

This chapter deals with modeling and simulation of a permanent magnet synchronous generator (PMSG)-based marine current turbine (MCT) with inter-turn faults. The generator is modeled in healthy and faulty conditions by using the (abc) reference frame and the (dq) reference frame. Indeed, the PMSG, installed under the sea (existence of the swell and wave), can be exposed to higher voltages and/or currents transited, which leads to the appearance of various faults. In this chapter, the faulty mode deals with the study of the stator inter-turn short-circuit faults in the PMSG. In fact, this fault presents a big problem because it can lead to the total degradation of the machine. Simulation results are carried out by using Matlab/Simulink environment.

Keywords: marine current turbine, permanent magnet synchronous generator, modeling, inter-turn faults, marine current turbine structure

1. Introduction

Oceans covering more than 70% of the earth, have long been acknowledged as a vast renewable energy source, such as thermal energy, wave energy, and marine tidal energy [1]. Indeed, the potential of electric power generation from marine tidal currents is very important; it has been shown that 48% of the European tidal resource is in the UK, 42% in France, and 8% in Ireland [2, 3].

Certainly, marine current turbine systems are exposed to environmental and functional constraints. Firstly, environmental constraints are due to the severe weather conditions because of the geographic location (installation under the sea and existence of harmonic current speeds caused by the swell and wave), second, functional constraints are due to the increase in power implies to higher voltages and/or currents transmitted. These constraints provide the degradation of performance of the various functional blocks of the system and its accelerated aging process, which leads to many faults essentially related to the blades, to the PMSG (short-circuit between turns, phases, and phase and neutral or faults in permanents magnets) and even to the rectifier (short-circuit faults, open-circuit faults, and intermittent gate misfiring faults).

Indeed, a permanent magnet synchronous generator has been chosen because of its advantages, such as high efficiency, compact structure, and the possibility to eliminate the gearbox, which reduce maintenance, and this is very favorable in terms of underwater application [4, 5].

However, the existence of inter-turn faults in the stator presents a big problem for some industrial applications car, if the fault is undetected, it can lead to other types of faults (short-circuit between phases or phase and neutral) and can virtually generate the total degradation of the stator winding, that is why the detection of this fault must be achieved earlier to prevent the spread of default to the other components of the system [6].

This chapter describes the modeling of the PMSG used in normal conditions and if an inter-turn short-circuit in stator winding has been presented [7].

This chapter is composed as follows: In Section 2, the MCT structure is given. In Section 3, modeling of PMSG in healthy conditions is presented. In Section 4, modeling of the PMSG in faulty conditions is given. In Section 5, simulation results are presented. The conclusion is given in Section 6.

2. Marine current turbine structure

As shown in **Figure 1**, the MCT structure is composed of marine turbine, a permanent magnet synchronous machine coupled to a DC bus through a PWM power rectifier.

2.1 Resource model

Tidal currents are proven by the effect of the moon and to a lesser degree, the sun, on the earth's surface. Since the moon is so much closer to the earth than the sun, its pull has more influence on the tides, the magnitude of the tide-generating force is about 68% of moon and 32% of sun. Indeed, the moon's gravitational pull forces the ocean to bulge outwards on opposite sides of the earth, which causes a rise in the water level in places that are aligned with the moon and a decrease in water levels halfway between those two places. This rise in water level is accompanied by a horizontal movement of water called the **tidal current**. **Figure 2** shows the simulation curve of marine current speed.

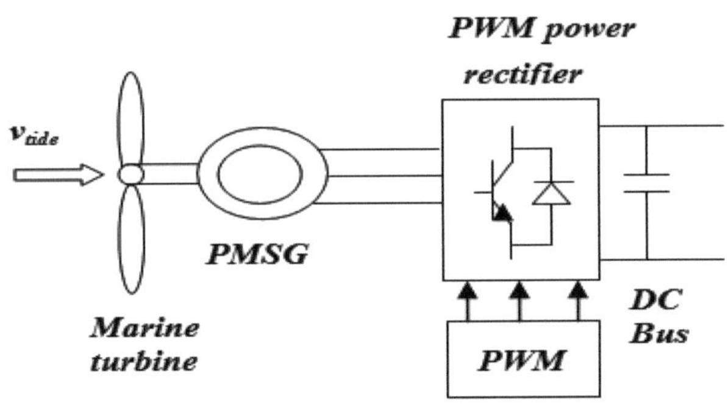

Figure 1.
Marine current turbine structure.

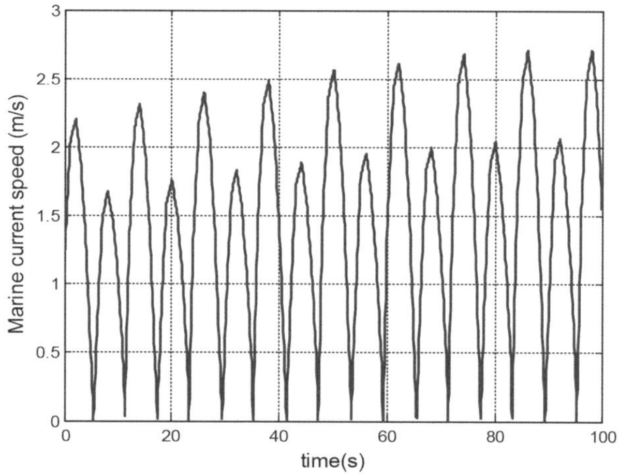

Figure 2.
Marine current speed.

2.2 Marine turbine rotor model

The conversion of kinetic energy into mechanical energy is achieved by using a marine turbine rotor. The mechanic power for a marine current turbine has the same dependence as that of a wind turbine and is given by the following equation [8, 9]:

$$P_m = \frac{1}{2} C_p(\lambda, \beta) \rho \pi r^2 v_t^3 \tag{1}$$

where ρ is the fluid density, r is the turbine radius, v_t is the tidal speed, and C_p is the power coefficient; it presents the percentage of mechanical power, which can be extracted from the fluid stream by the turbine.

For typical MCTs, the maximum value of C_p for normal operation is estimated to be in the range of [0.35–0.5] [10]. Indeed, based on the experimental results and for a given turbine, the C_p can be approximated as an expression of the blade pitch angle β and the tip speed ratio λ and Ref. [11]. **Figure 3** illustrates the C_p curve for simulations.

2.3 Generator model

To model the PMSG, we will use two reference frame: the (abc) reference frame and the (dq) reference frame. The first one is the most used in the literature in the case of faulty conditions. The second is the (dq) reference frame without neglecting the zero-sequence component. Indeed, in the presence of short-circuit; this zero-sequence component is not zero.

The modeling of the PMSG will be developed in the next section in both cases: healthy and faulty conditions.

2.4 PWM power rectifier model

The PWM power rectifier uses the three-phase voltage sources provided by the PMSG.

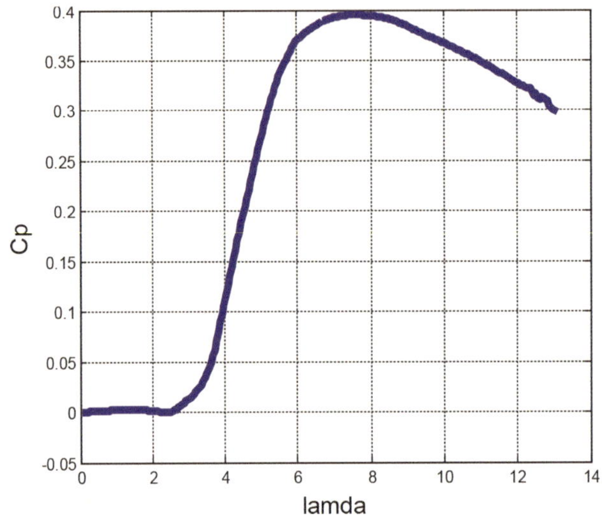

Figure 3.
Power coefficient curve C_p.

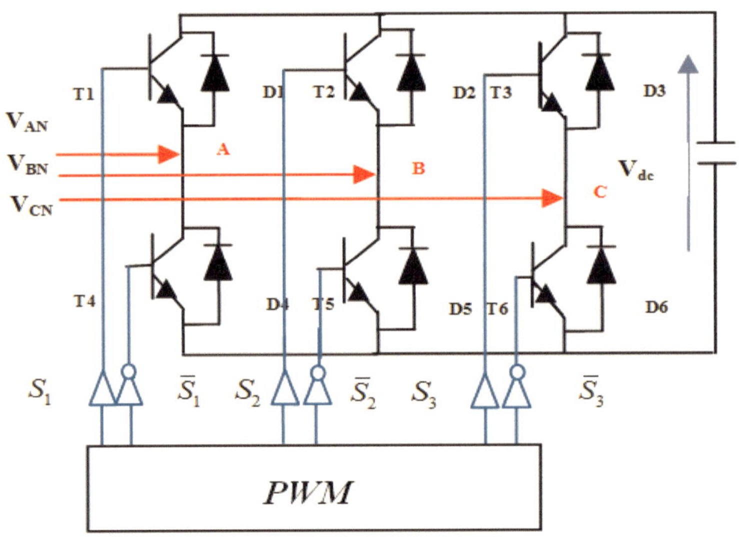

Figure 4.
PWM power rectifier topology.

As shown in **Figure 4**, this converter is composed of three legs, each leg features two semiconductor switches (T_k, T_{k+3} k = 1, 2, 3) with antiparallel connected freewheeling diodes (D_k, D_{k+3}). The switches of the same leg are controlled by a PWM bloc in the form of a logic control signal S_k (k = 1, 2, 3) also known as gate signals, it is defined by:

$$S_k = \begin{cases} 1 \, if \ T_k on \ and \ T_{k+3} \ off \\ 0 \, if \ T_{k+3} on \ and \ T_k \ off \end{cases} \quad (2)$$

3. Generator model in healthy conditions

3.1 abc-frame

The PMSG is presented in abc-frame as shown in **Figure 5**.

Dynamic modeling can be expressed as follows:

$$[v_s] = [R_s][i_s] + \frac{d[\psi_s]}{dt} \tag{3}$$

with:

$$[v_s] = [v_a \ v_b \ v_c]^T \tag{4}$$

$$[i_s] = [i_a \quad i_b \quad i_c]^T \tag{5}$$

$$[\psi_s] = [\psi_a \quad \psi_b \quad \psi_c]^T \tag{6}$$

$$[R_s] = \begin{bmatrix} R_s & 0 & 0 \\ 0 & R_s & 0 \\ 0 & 0 & R_s \end{bmatrix} \tag{7}$$

where v_s is the stator voltage, i_s is the stator current, R_s is the stator resistance, and ψ_s is the stator magnetic flux.

The stator magnetic flux is given by:

$$[\psi_s] = [L_s][i_s] + [\psi_{ms}] \tag{8}$$

with:

$$[L_s] = \begin{bmatrix} L & M & M \\ M & L & M \\ M & M & L \end{bmatrix} \tag{9}$$

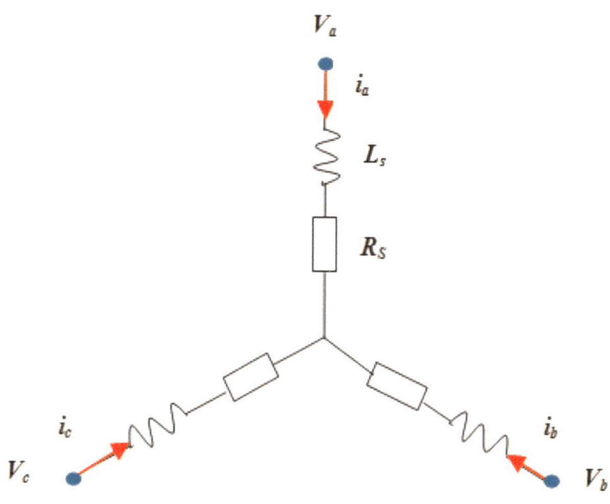

Figure 5.
PMSG model in abc-frame.

$$[\psi_{ms}] = [\psi_{ma} \quad \psi_{mb} \quad \psi_{mc}]^T = \psi_m \left[\sin(\theta) \quad \sin\left(\theta - \frac{2\pi}{3}\right) \quad \sin\left(\theta - \frac{4\pi}{3}\right) \right]^T \tag{10}$$

where L is the self-inductance, M is the mutual inductance, and ψ_m is the permanent magnet flux.

For a balanced system:

$$i_a + i_b + i_c = 0 \tag{11}$$

The inductance matrix becomes:

$$[L_s] = \begin{bmatrix} L-M & 0 & 0 \\ 0 & L-M & 0 \\ 0 & 0 & L-M \end{bmatrix} = \begin{bmatrix} L_s & 0 & 0 \\ 0 & L_s & 0 \\ 0 & 0 & L_s \end{bmatrix} \tag{12}$$

3.2 dq-frame

Park's transformation allows the conversion of a three-phase winding to a two-phase winding. Its expression is given as follows:

$$\begin{bmatrix} \boldsymbol{X_d} \\ \boldsymbol{X_q} \\ \boldsymbol{X_0} \end{bmatrix} = [\boldsymbol{P(\theta)}] \begin{bmatrix} \boldsymbol{X_a} \\ \boldsymbol{X_b} \\ \boldsymbol{X_c} \end{bmatrix} \tag{13}$$

where

$$[P(\theta)] = \sqrt{\frac{2}{3}} \begin{bmatrix} cos\theta & \cos\left(\theta - \frac{2\pi}{3}\right) & \cos\left(\theta - \frac{4\pi}{3}\right) \\ -sin\theta & -\sin\left(\theta - \frac{2\pi}{3}\right) & -\sin\left(\theta - \frac{4\pi}{3}\right) \\ \frac{1}{\sqrt{2}} & \frac{1}{\sqrt{2}} & \frac{1}{\sqrt{2}} \end{bmatrix} \tag{14}$$

X can be voltage, current, or flux.

The factor $\sqrt{\frac{2}{3}}$ has been introduced to ensure the condition of invariance power.

The application of Park's transformation to the stator voltage equations allows the following relations:

$$\begin{cases} v_d = R_s i_d + \dfrac{d[\psi_d]}{dt} - p\Omega\psi_q \\ v_q = R_s i_q + \dfrac{d[\psi_q]}{dt} + p\Omega\psi_d \end{cases} \tag{15}$$

where

$$\begin{cases} \psi_d = L_s i_d + \psi_m \\ \psi_q = L_s i_q \end{cases} \tag{16}$$

The machine state space form is written as follows [12–14]:

$$\begin{cases} \frac{di_d}{dt} = -\frac{R_s}{L_s} i_d + p\Omega i_q + \frac{v_d}{L_s} \\ \frac{di_q}{dt} = -\frac{R_s}{L_s} i_q - p\Omega i_q - \frac{p\psi_m}{L_s}\Omega + \frac{v_q}{L_s} \\ \frac{d\Omega}{dt} = \frac{p}{J} T_m - \frac{p}{J} T_{em} - \frac{p}{J} f\Omega \\ T_{em} = \frac{3}{2} p\psi_m i_q \end{cases} \tag{17}$$

where p is the pair pole number, Ω is the turbine speed, T_m is the mechanical torque, T_{em} is the electromagnetic torque, f is the viscosity coefficient, and J is the turbine and the PMSG inertia.

4. Generator model in faulty conditions

4.1 abc-frame

In this section and as shown in **Figure 6**, an inter-turn short-circuit is occurred in phase "a."

We define:

$$\mu = \frac{N_{sh}}{N_s} \tag{18}$$

where.

N_{sh}: the number of shorted phase turns.

N_s: the total number of phase turns.

If μ = 0, the faulty model will reduce to a healthy model.

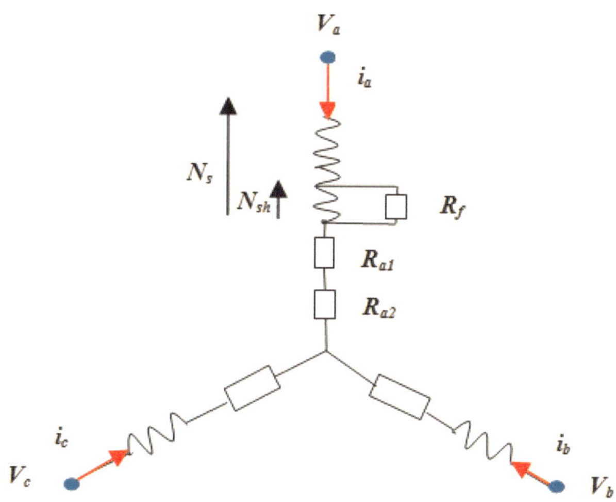

Figure 6.
Three-phase stators winding with turn fault in phase "a."

R_f is the fictitious resistance across the shorted winding to limit the short-circuit current [7], it defined by:

$$R_f = \begin{cases} \infty \text{ in healthy conditions} \\ 0 \text{ in faulty conditions} \end{cases} \quad (19)$$

The stator equations model is modified as follows:

$$[v_s] = \begin{bmatrix} v_{a1} \, v_{a2} & v_b & v_c \end{bmatrix}^T \quad (20)$$

$$[i_s] = \begin{bmatrix} i_a \, i_a - i_f & i_b & i_c \end{bmatrix}^T \quad (21)$$

where v_{a1} is the voltage across the non-shorted winding part, v_{a2} is the voltage across the shorted winding part, and i_f is the additional current spawns by the short-circuit.

The resistance matrix is expressed as follows:

$$[R_s] = \begin{bmatrix} R_{a1} & 0 & 0 & 0 \\ R_{a2} & 0 & 0 & 0 \\ R_s & 0 & 0 & 0 \\ R_s & 0 & 0 & 0 \end{bmatrix} \quad (22)$$

where

$$\begin{cases} R_{a1} = (1-\mu)R_s \\ R_{a2} = \mu R_s \end{cases} \quad (23)$$

The addition of rows 1 and 2 of Eq. (20) allows us to obtain new machine equations as follows [15]:

$$[v_{sf}] = [R_s][i_s] + \frac{d[\psi_s]}{dt} - R_{a2} i_f - \frac{d[\psi_f]}{dt} \quad (24)$$

where ψ_f is the short-circuit fault flux defined by:

$$[\psi_f] = [L_f] i_f \quad (25)$$

$$[L_f] = [L_{a2} + M_{a1a2} \quad M_{a2b} \quad M_{a2c}]^T \quad (26)$$

We note that Eq. (24) is composed of two parts: a healthy part and a faulty part, caused by the short-circuit fault.

To solve this equation, we must add the expression of the voltage across the shorted winding part defined as follows:

$$v_{a2} = R_{a2}(i_a - i_f) + L_{a2}\frac{d(i_a - i_f)}{dt} + M_{a1a2}\frac{di_a}{dt} + M_{a2b}\frac{di_b}{dt} + M_{a2c}\frac{di_c}{dt} + \frac{d\psi_{mb}}{dt} = R_f i_f \quad (27)$$

Because of inter-turn faults, the size of the system becomes 4. The final model of the PMSG in the abc-frame is presented as follows:

$$\begin{bmatrix} v_a \\ v_b \\ v_c \\ 0 \end{bmatrix} = \begin{bmatrix} R_s & 0 & 0 & -R_{a2} \\ 0 & R_s & 0 & 0 \\ 0 & 0 & R_s & 0 \\ -R_{a2} & 0 & 0 & R_{a2}+R_f \end{bmatrix} \begin{bmatrix} i_a \\ i_b \\ i_c \\ i_f \end{bmatrix} + \begin{bmatrix} L_s & 0 & 0 & -L_{a2}-M_{a1a2} \\ 0 & L_s & 0 & -M_{a2b} \\ 0 & 0 & L_s & -M_{a2c} \\ -L_{a2}-M_{a1a2} & -M_{a2b} & -M_{a2c} & L_{a2} \end{bmatrix} \frac{d}{dt} \begin{bmatrix} i_a \\ i_b \\ i_c \\ i_f \end{bmatrix} + \frac{d}{dt} \begin{bmatrix} \psi_{ma} \\ \psi_{mb} \\ \psi_{mc} \\ -\psi_{ma2} \end{bmatrix} \tag{28}$$

4.2 dq-frame

In faulty conditions, dynamic modeling in dq-frame can be given by this expression:

$$\left[v_{dqf}\right] = [R_s]\left[i_{dq}\right] + \frac{d\left[\psi_{dq}\right]}{dt} \pm \omega\left[\psi_{dq}\right] - \sqrt{\frac{2}{3}}R_{a2}i_f - \frac{d\left[\psi_{dqf}\right]}{dt} \tag{29}$$

where

$$\left[\psi_{dqf}\right] = \left[M_{fd} \quad M_{fq}\right] i_f \tag{30}$$

$$M_{fd} = \sqrt{\frac{2}{3}}\left(L_{a2} + M_{a1a2} - \frac{M_{a2b}+M_{a2c}}{2}\right) \tag{31}$$

$$M_{fq} = \frac{1}{\sqrt{2}}\left(M_{a2b} + M_{a2c}\right) \tag{32}$$

The final representation is given as follows:

$$\begin{bmatrix} v_d \\ v_q \\ 0 \end{bmatrix} = \begin{bmatrix} R_s & 0 & -R'_{a2} \\ 0 & R_s & 0 \\ -R'_{a2} & 0 & R'_f \end{bmatrix} \begin{bmatrix} i_d \\ i_q \\ i_f \end{bmatrix} + \begin{bmatrix} L_s & 0 & M_{fd} \\ 0 & L_s & M_{fq} \\ M_{fd} & M_{fq} & L_s \end{bmatrix} \frac{d}{dt} \begin{bmatrix} i_d \\ i_q \\ i_f \end{bmatrix} + \frac{d}{dt} \begin{bmatrix} \psi_d \\ \psi_d \\ -\psi_{ma2} \end{bmatrix} \tag{33}$$

where

$$\begin{cases} R'_{a2} = \sqrt{\frac{2}{3}}R_{a2} \\ R'_f = R_{a2} + R_f \end{cases} \tag{34}$$

5. Simulation results

In this section, simulation software has been set up by using the Matlab/Simulink environment. An inter-turn short fault is applied to phase "*a*" at t = 0.3 s.

The parameters used in simulation tests are displayed in **Table 1**.

The waveforms given in **Figures 7–9** show the three-phase currents, when (R_f = 1 Ω and μ = 0.1).

We note that the current magnitude in phase "*a*" increases after the application of the fault. Whereas, in phase "*b*" and phase "*c*," the rise of current is very slight.

Figure 10 shows the current flowing through the fictitious resistance branch; we note that this current drops from zero to a sinusoidal form.

The speed and the torque are given in **Figures 11** and **12**. After the fault instant; we note that there is an appearance of important corrugations.

To study the impact of default between turns on the behavior of the machine, firstly, the simulation is done for different values of the default resistance, then, for the different number of shorted turns.

The three-phase currents and the current in the fictitious resistance are given in **Figures 13–15**. When the fault resistance decreases, the magnitude of the default

MCT parameters	
Turbine blade radius	0.87 m
Number of blades	3
Fluid density	1027.68 Kg/m^3
Generator stator resistance	0.173 mΩ
Generator d-axis reference	0.085 mH
Generator q-axis reference	0.951 mH
Permanent magnet flux	0.112 Wb
Generator inertia	0.0048 $kg.m^2$
Viscosity coefficient	8.5 10^{-3} Nm/s

Table 1.
MCT parameters.

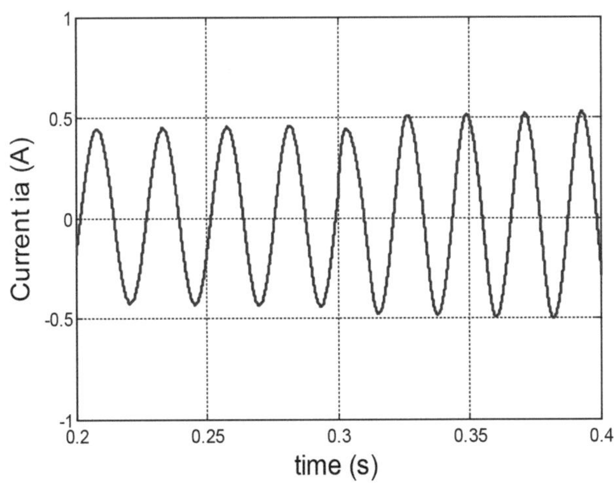

Figure 7.
Current i_a *before and after fault.*

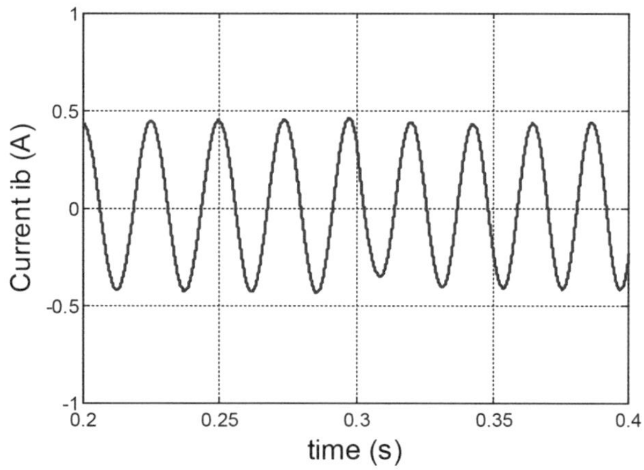

Figure 8.
Current i_b *before and after fault.*

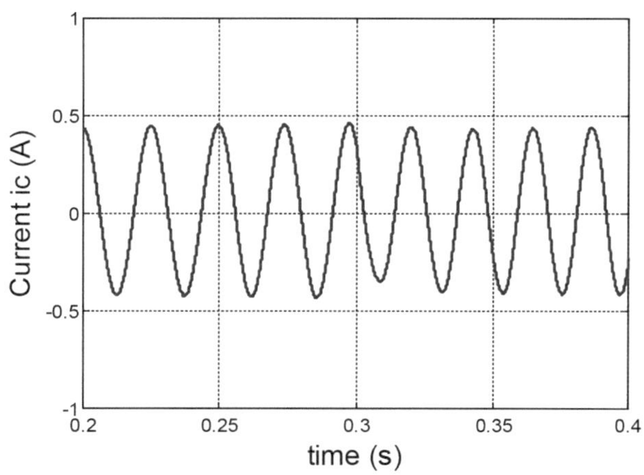

Figure 9.
Current i_c *before and after fault.*

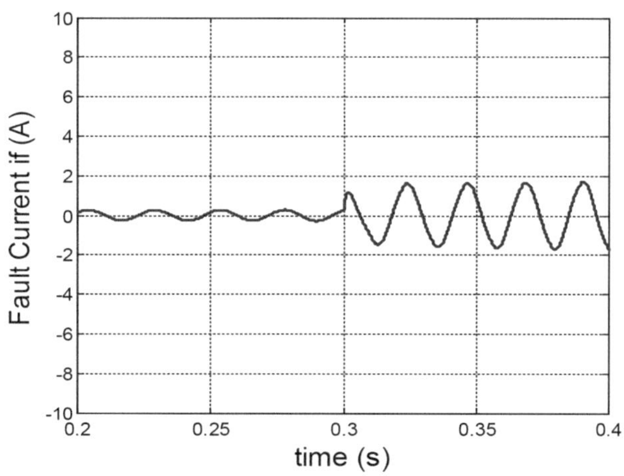

Figure 10.
Fault current i_f.

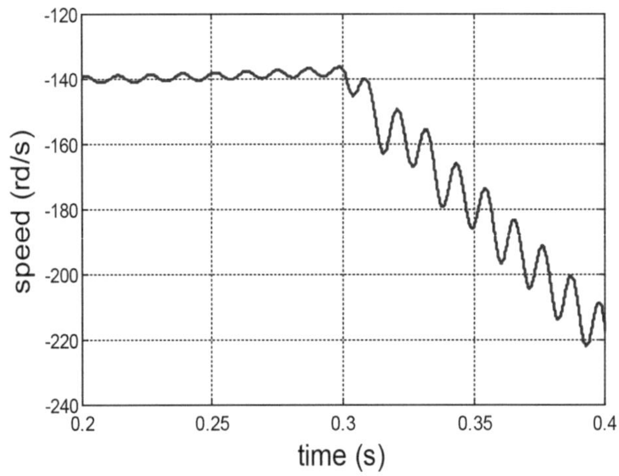

Figure 11.
Speed curve before and after fault.

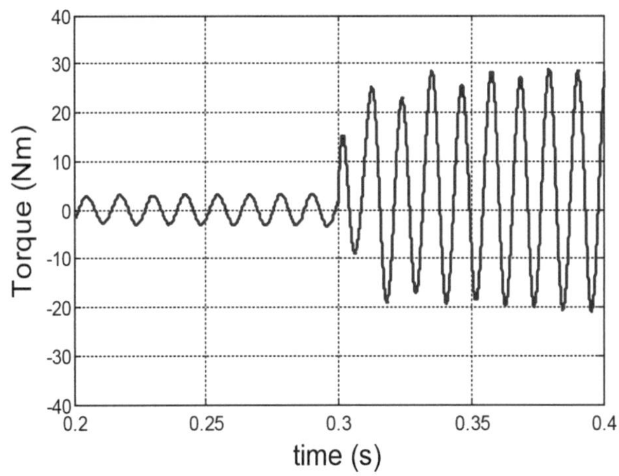

Figure 12.
Torque curve before and after fault.

phase current becomes more elevated than the other healthy phase currents. Also, the current in the fictitious resistance increases.

Figure 16 shows the evolution of the speed for a different number of shorted turns. We note that because of the increased number of shorted turns, the corrugations become increasingly important and the machine starts to unhook.

6. Conclusion

This chapter presents the modeling and simulation of a permanent magnet synchronous generator with inter-turn short circuits on the stator. The equations of the machine in healthy and faulty conditions are presented in both (abc) reference frame

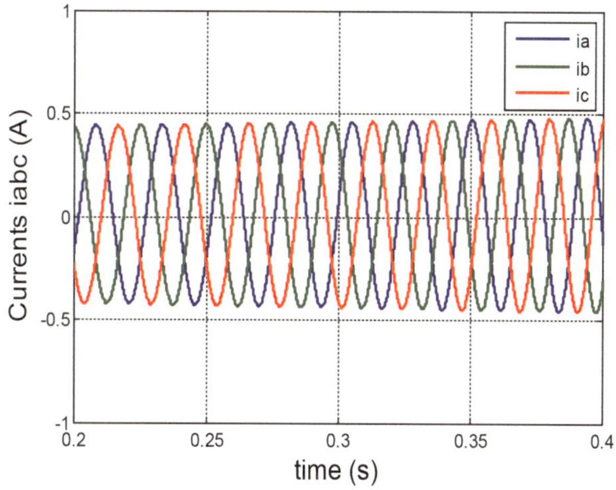

(a)

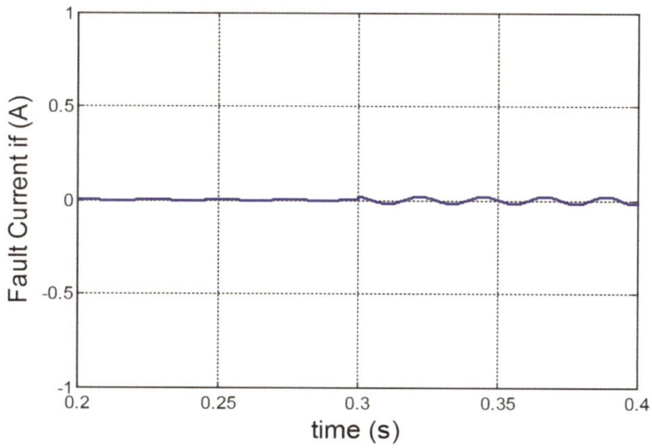

(b)

Figure 13.
(a) Phase currents i_{abc}, (b) fault current i_f (R_f = 10 Ω).

and (dq) reference frame. The main focus has been the study of the impact of this fault on the three-phase currents, speed, and torque. In addition, the impact of this fault is related to the default resistance and the number of shorted turns. The proposed system has been performed by using Matlab/Simulink.

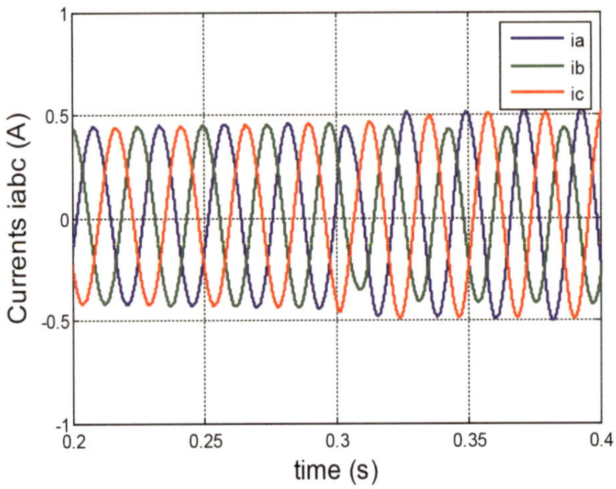

(a)

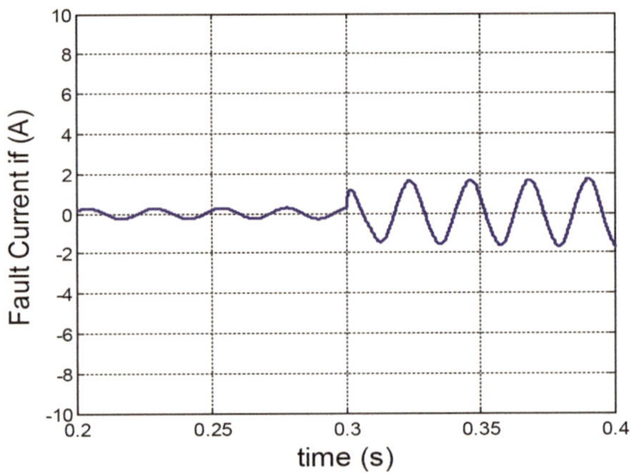

(b)

Figure 14.
(a) Phase currents i_{abc}, (b) Fault current i_f (R_f = 1 Ω).

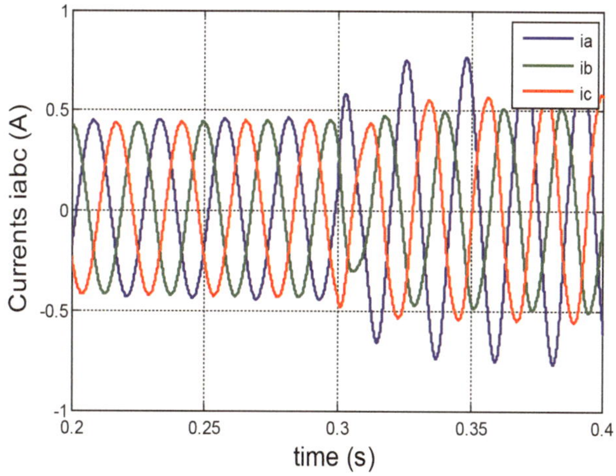

(a)

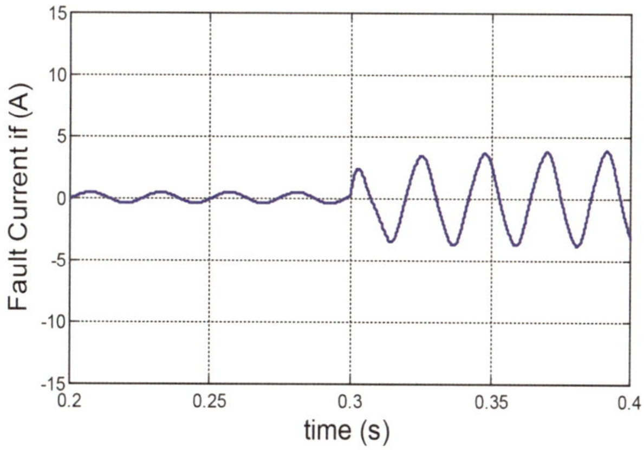

(b)

Figure 15.
(a) Phase currents i_{abc}, (b) Fault current i_f (R_f = 0.01 Ω).

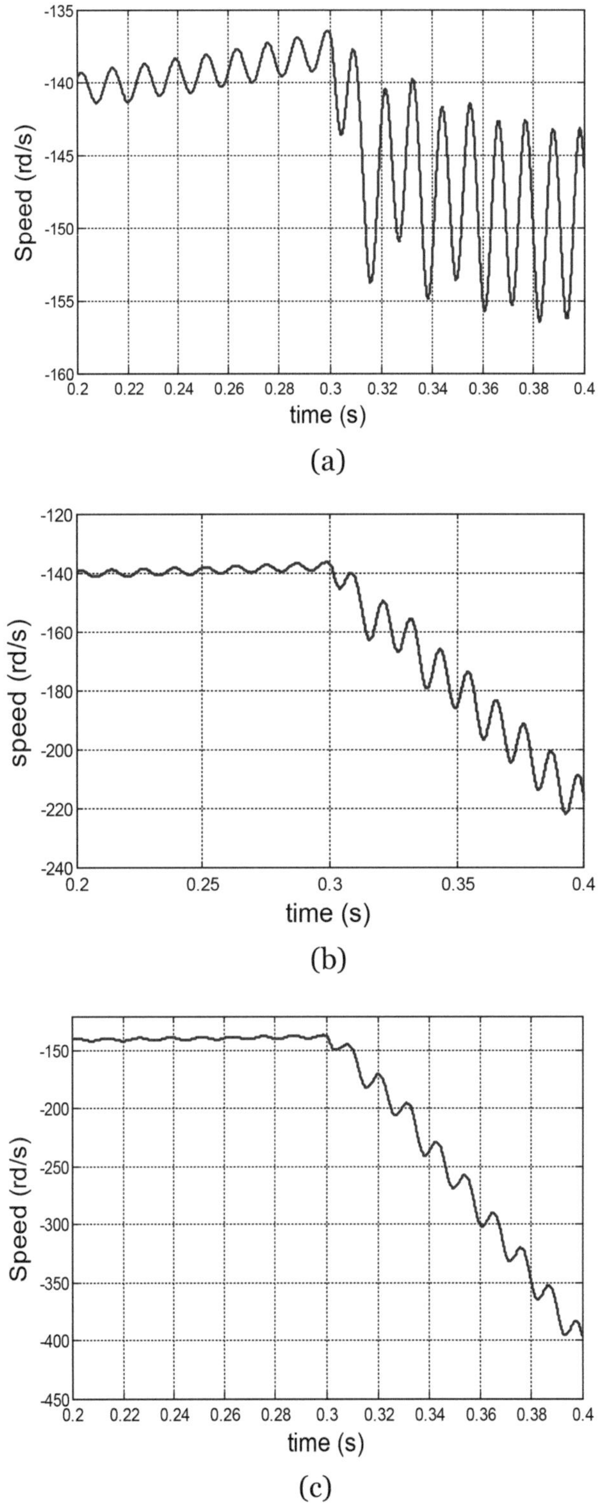

Figure 16.
Speed curve (a) μ = 8%, (b) μ = 10%, and (c) μ = 13%.

Author details

Sana Toumi[1,2]*, Mohamed Benbouzid[2,3] and Mohamed Faouzi Mimouni[1]

1 Laboratory of Automation, Electrical Systems and Environment (LASEE), University of Monastir, Monastir, Tunisia

2 University of Brest, Brest, France

3 Shanghai Maritime University, Shanghai, China

*Address all correspondence to: sanaatoumii@gmail.com

References

[1] Toumi S, Amirat Y, Elbouchikhi E, Trabelsi M, Mimouni MF, Benbouzid MEH. Second-order sliding mode for marine current turbine fault-tolerant control. In: the IEEE International Conference on Control Engineering & Information technology (CEIT'04); 16-18 December 2016; Hammamet. Tunisia: IEEE; 2016. pp. 1-6

[2] Toumi S, Amirat Y, Trabelsi M, Elbouchikhi E, Mimouni MF, Benbouzid MEH. Backstepping control of a PMSG-based marine current turbine system under faulty conditions. In: Proceedings of the IEEE International Renewable Energy Congres (IREC'09); 20–22 March 2018, Hammamet. Tunisia: IEEE; 2018. pp. 1-6

[3] Toumi S, Benelghali S, Trabelsi M, Elbouchikhi E, Amirat Y, Benbouzid MEH, et al. Modeling and simulation of a PMSG-based marine current turbine system under faulty rectifier conditions. Electric Power Components and Systems. 2017;**45**: 715-725

[4] Conroy JF, Watson R. Frequency response capability of full converter wind turbine generators in comparison to conventional generation. IEEE Transactions Power Systems. 2008;**23**: 649-656

[5] Zhou Z, Scuiller F, Charpentier JF, Benbouzid MEH, Tang T. Grid-connected marine current generation system power smoothing control using supercapacitors. In: Proceedings of the IEEE Annual Conference on IEEE Industrial Electronics Society (IECON'38); 25–28 October 2012, Montreal. Canada: IEEE; 2012. pp. 1-6

[6] Bachir S, Tnani S, Trigeassou JC, Champenois G. Diagnosis by parameter estimation of stator and rotor faults occurring in induction machines. IEEE Transactions Industrial Electronics. 2006;**53**:963-973

[7] Farouq JA, Raminosoa T, Djerdir A, Miraoui A. Modelling and Simulation of stator winding inter-turn faults in permanent magnet synchronous motors. COMPEL Journal. 2008;**27**:887-896

[8] Toumi S, Amirat Y, Elbouchikhi E, Trabelsi M, Benbouzid MEH, Mimouni MF. A comparison of fault-tolerant control strategies for a PMSG-based marine current turbine system under converter faulty conditions. Journal of Electrical Systems. 2017;**13**: 472-488

[9] Benelghali S, Balme R, Le Saux K, Benbouzid MEH, Charpentier JF, Hauville F. A simulation model for the evaluation of the electrical power potential harnessed by a marine current turbine. IEEE Journal on Oceanic Engineering. 2007;**32**:786-797

[10] Toumi S, Benelghali S, Trabelsi M, Elbouchikhi E, Benbouzid MEH, Mimouni MF. Robustness analysis and evaluation of a PMSG-based marine current turbine system under faulty conditions. In: Proceedings of the IEEE International Conference on Sciences and Techniques of Automatic Control and and Computer Engineering (STA'15); 21–23 December 2014; Hammamet. Tunisia: IEEE; 2014. pp. 631-636

[11] Toumi S, Amirat Y, Elbouchikhi E, Trabelsi M, Benbouzid MEH, Mimouni MF. A simplified mathematical approach for magnet defects modeling in PMSG-based marine current turbine. In: Proceedings of the IEEE International Conference on Sciences and Techniques of Automatic Control and Computer

Engineering (STA'17); 19–21 December 2016; Sousse. Tunisia: IEEE; 2016. pp. 552-557

[12] Errami Y, Maaroufi M, Ouassaid M. Modelling and Control Strategy of PMSG Based Variable Speed Wind Energy Conversion System. In: Proceedings of the IEEE International Conference on Multimedia Computing and Systems (ICMCS); 7–9 April 2011; Quarzazate. Morocco: IEEE; 2011. pp. 1-6

[13] Poddar G, Joseph A, Unnikrishnan AK. Sensorless variable-speed controller for existing fixed-speed wind power generator with unity-power-factor operation. IEEE Transactions Industrial Electronics. 2003;**50**:1007-1015

[14] Weizheng Y, Woo K, Ruijie Z, Wei G, Yue W. Analyze of current control strategy based on vector control for permanent-magnet synchronous generator in wind power system. In: Proceedings of the IEEE International Power Electronics and Motion Control Conference (IPEMC'6); 17–20 May 2009; Wuhan. China: IEEE; 2009. pp. 2209-2212

[15] Tallam RM, Habetler TG, Harley RG. Transient model for induction machines with stator winding turn faults. IEEE Transactions Industrial Applications. 2002;**38**:623-627

Chapter 10

Sensorless Speed Observer for Industrial Drives Based Induction Motors with Low Complexity

Sherif Dabour, Mohamed Hussien, Ahmed A. Aboushady and Mohamed Emad Farrag

Abstract

This chapter presents a simple method to efficiently predict the rotor speed for a sensorless vector control technique applied to induction motors (IMs). The motor is supplied by a Simplified Split-Source Inverter (S^3I), which provides dc-boosting and ac-inversion processes during input voltage sag. It also has a wider modulation range and a lower harmonic content than conventional boosting inverters. With this contribution, it is possible to efficiently estimate the rotor position directly without needing a PI controller with fluctuated supply voltage. The proposed strategy can be divided into three parts. The first uses a dual-loop controller to obtain the reference DC-boosted voltage of the SSI and regulate the input current. The second is the suggested observer for speed detection, which is derived from the principles of phase axis relations of the adopted machine currents and the indirect rotor flux orientation control (IRFOC) approach. With a newly developed space vector modulation, the third part will generate the switching pulses of the inverter switches. A complete analysis has been conducted to ensure the observability of the proposed technique. A series of PLECS simulations were conducted to verify the concept. The obtained results validate the proposed strategy with the S^3I topology.

Keywords: induction motor, split-source inverter, sensorless observer for speed detection, vector control technique, indirect field-oriented control (IRFOC)

1. Introduction

Due to their low cost, high power density, and ability to withstand harsh environments, induction motors are the most popular motors in the industry [1, 2]. If this motor is directly connected to the grid, it operates at speeds determined by the line frequency and the number of poles. Due to the nonlinear relationship between motor current and torque, speed control is much more complex than DC motors. During the past few years, there have been numerous applications for induction motors, including pumps, blowers, and industrial applications. Although this control strategy is easy to implement and utilizes inexpensive hardware, it has poor performance and slow response times. Nevertheless, modern variable speed drive applications, such as electric vehicles (EVs), require high performance and fast dynamic response control systems.

To improve the dynamic responses of the electric drive systems, vector control strategies, such as the Field-oriented Control (FOC) method, are proposed [3–5]. According to this control method, flux is controlled by the amount of current generated on the d-axis, while the current component on the q-axis controls torque. Based on this approach, the induction motor mimics the dynamic behavior of a DC motor. The advantages of FOC include a wider speed range, better dynamic response, precise speed and position control, and field weakening (movement beyond base speed). In this control method, it is possible to directly measure the position of the field or the motor speed using the hall effect sensor. However, this method increases the system cost and the need for maintenance.

Low-cost and highly reliable controlled IM drives without mechanical sensors at the motor shaft have been demonstrated in the literature. It's called sensorless vector control algorithms. In these algorithms, to replace the sensor, the stator voltages and currents at the motor terminals are measured to determine the rotor speed or position [3–5]. The dynamic equations of induction motors can be used to estimate several quantities, such as magnetic flux and frequency. However, these quantities often require expensive or difficult-to-implement physical sensors. This class only requires current and voltage sensors. Accurate knowledge of machine parameters and open equations for estimates is necessary. On the other hand, closed-loop observatories rely on flexible techniques that require more computational power [6–18].

In general, a conventional voltage source inverter (VSI) can only achieve an output voltage that is limited by the input voltage level of the DC supply [19, 20]. Therefore, the new topology overcomes the performance-enhancing potential of DC-AC power converters, such as the Z-Source Topology Converter (ZSI). However, there are some issues, such as the lack of input current continuity and the use of four passive elements. Therefore, S^3I (Simplified Split Source Inverter) topology can be used instead of ZSI, as shown in **Figure 1**, which handles the benefits of continuous input current and the reduced need for passive components [21]. Due to their advantages, the SSI is recently considered in different applications and control techniques, such as model predictive control [22], virtual synchronous control [23], dual motor drive [24], multilevel converter topologies [25] and multiphase machine drives [26].

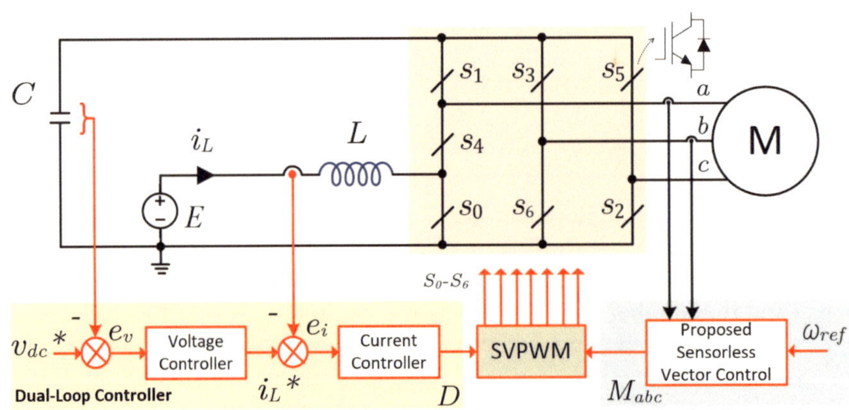

Figure 1.
Simplified Split-source inverter feeding three-phase induction motor.

This chapter aims to present a simple method for efficiently predicting the induction motor speed, which is supplied by S^3I directly without a speed sensor. Basically, there are three parts to the control strategy. In the first part, the dc-link voltage v_{dc} and the supply current i_L, shown in **Figure 1**, are controlled by a dual-loop controller. In the second part, we propose a new speed observer derived from the principles of phase axis relationships of the adopted machine currents and indirect rotor flux orientation control (IRFOC). The third part will be used to generate the gating pulses of S^3I with a modified SVPWM approach. Finally, the concept behind this chapter was verified using simulations.

2. Dynamic model of IM

The voltage and current relations for three-phase IM in the dq-axes are described as follows:

$$\begin{cases} v_{q_s} = r_s i_{q_s} + \dfrac{d}{dt}\lambda_{q_s} + \omega\lambda_{d_s} \\ v_{d_s} = r_s i_{d_s} + \dfrac{d}{dt}\lambda_{d_s} - \omega\lambda_{q_s} \\ 0 = r_r i_{q_r} + \dfrac{d}{dt}\lambda_{q_r} + (\omega - \omega_r)\,\lambda_{d_r} \\ 0 = r_r i_{d_r} + \dfrac{d}{dt}\lambda_{d_r} - (\omega - \omega_r)\,\lambda_{q_r} \end{cases} \quad v_{q_s} = r_s i_{q_s} + \frac{d\lambda_{q_s}}{dt} + \omega\lambda_{d_s} \tag{1}$$

where,

$$\begin{cases} \omega = \dfrac{P}{2}\,\omega_f \\ \omega_r = \dfrac{P}{2}\,\omega_{rm} \end{cases} \tag{2}$$

The associated relations of flux linkage are given as follows

$$\begin{cases} \lambda_{q_s} = L_s i_{q_s} + L_m i_{q_r} \\ \lambda_{d_s} = L_s i_{d_s} + L_m i_{d_r} \\ \lambda_{q_r} = L_r i_{q_r} + L_m i_{q_s} \\ \lambda_{d_r} = L_r i_{d_r} + L_m i_{d_s} \end{cases} \tag{3}$$

where,

$$\begin{cases} L_s = L_{ls} + L_m \\ L_r = L_{lr} + L_m \end{cases} \tag{4}$$

The corresponding electromagnetic torque can be obtained as follows

$$T_e = \left(\frac{3}{2}\right)\left(\frac{P}{2}\right)\left(\frac{L_m}{L_r}\right)\left(\lambda_{d_r} i_{q_s} - \lambda_{q_r} i_{d_s}\right) \tag{5}$$

Furthermore, the overall torque expression can be realized by

$$T_e = T_L + J\frac{d\omega_{rm}}{dt} + B\,\omega_{rm} \tag{6}$$

3. Proposed sensorless vector-control strategy

In general, vector control strategies are classified into direct field-oriented control (DFOC) and indirect field-oriented control (IRFOC) [3–5]. This chapter concerns the IRFOC technique. According to the field orientation principle of the three-phase IM, the magnetic flux and motor torque can be controlled independently by injecting appropriate stator currents or voltages [4]. The dynamic equations of the three-phase IM, which are presented by (1)–(6), can be simplified under certain conditions. For rotor-driven control, the rotor flux is aligned with the excitation frame ($\lambda_{d_r} = \lambda_r$) and the expression is simplified so that λ_{q_r} equals zero ($\lambda_{q_r} = 0$). Since the resulting motor's torque and flux are decoupled. In this way, the IM can be controlled like a DC motor [2]. In this technique, the synchronous speed required to maintain the orthogonal orientation of vectors λ_r and i_r in field-oriented control (FOC) is denoted by $\omega_e{}^*$. For the proposed orientation of the rotor winding field, the IM phase-axis relationship is shown in **Figure 2**.

The required slip speed ω_r and the desired current $i_s{}^*$ to produce the torque $T_e{}^*$ and the desired magnetic flux $\lambda_r{}^*$ are calculated from the motor equations and the orientation conditions of the magnetic field. Manipulating the equilibrium equations, we obtain the following equation, which represents an indirect vector controller:

$$i_{q_s}{}^* = \left(\frac{2}{3}\right)\left(\frac{2}{P}\right)\left(\frac{L_r}{L_m}\right)\frac{T_e{}^*}{\lambda_r} \tag{7}$$

$$i_{d_s}{}^* = \frac{\lambda_r{}^*}{L_m} \tag{8}$$

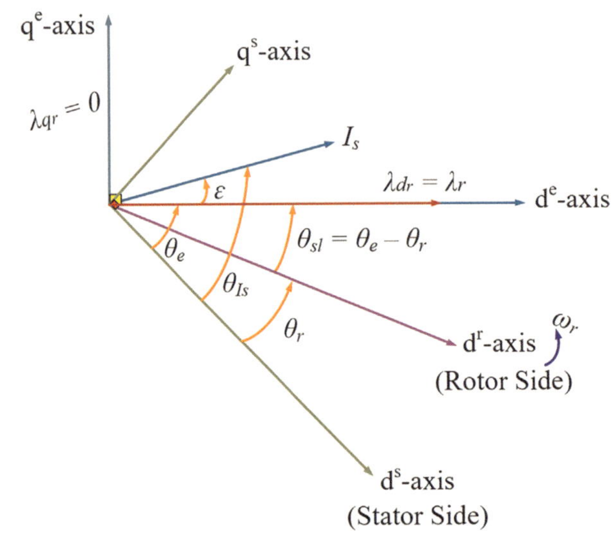

Figure 2.
Phase-axis relationship of the IM for the proposed control strategy.

Furthermore, the angular reference speed $\omega_e{}^*$ can be obtained from the machine rotor speed ω_r, and the slip speed ω_{sl} as follows

$$\omega_e{}^* = \omega_{sl} + \omega_r \tag{9}$$

Note that the relevant frame angle θ_e can be found directly by integrating $\omega_e{}^*$. In the indirect FOC (IRFOC) technique, the slip speed, ω_{sl} is determined from [2], as follows

$$\omega_{sl} = \frac{L_m\, r_r}{\lambda_r\, L_r}\, i_{q_s} \tag{10}$$

Based on the principles of the introduced IRFOC, the real flux of the rotor λ_r is calculated as:

$$\frac{L_r}{r_r}\frac{d\lambda_r}{dt} + \lambda_r = L_m i_{d_s} \tag{11}$$

The stator current component corresponding to the dq-axis can be determined from the actual three-phase currents using the desired frame angle (frame angle, $\theta_f = \theta_e$) with a suitable conversion (given K_{tr}):

$$K_{tr} = \frac{2}{3}\begin{bmatrix} cos\,\theta_f & cos\left(\theta_f - \frac{2\pi}{3}\right) & cos\left(\theta_f + \frac{2\pi}{3}\right) \\ -sin\,\theta_f & -sin\left(\theta_f - \frac{2\pi}{3}\right) & -sin\left(\theta_f + \frac{2\pi}{3}\right) \end{bmatrix} \tag{12}$$

It is worth noting that the vector control algorithm for the desired orientation depends on the rotor speed signal. However, the necessary speed sensor increases the overall system cost and requires more maintenance. Therefore, it is recommended to implement speed control based magnetic field orientation using a more efficient sensorless rotor speed monitor. Therefore, this section explores a simple method to detect the speed of the IM without an additional observer. The block diagram of the proposed sensorless control strategy is shown in **Figure 3**. The main control procedure applied can be illustrated as follows:

First, the phase angle, θ_{I_s} and the current space vector of the stator side, i_s are determined from the actual three-phase signals as follows

$$\begin{cases} i_s = k \cdot i_{abc_s} \\ \theta_{I_s} = angle\ (\ i_s\ ) \end{cases} \tag{13}$$

where $k = \left(2/3\ \ 2/3e^{j\frac{2\pi}{3}}\ \ 2/3e^{j\frac{4\pi}{3}}\right)$

In addition, the slip angle θ_{sl} can be obtained directly from the relevant slip angle velocity integration process using (10). Therefore, the corresponding amount of the expected rotor current on the dq axis is given using a transformation matrix, where K_{tr} is the angle of the indicated magnetic flux direction ($\theta_f = \theta_{sl}$).

Then, the corresponding shaft stator current dq can be determined using the IRFOC ($\lambda_{q_r} = 0$) in (3) and (11). Therefore, the relevant phase angle is expressed as

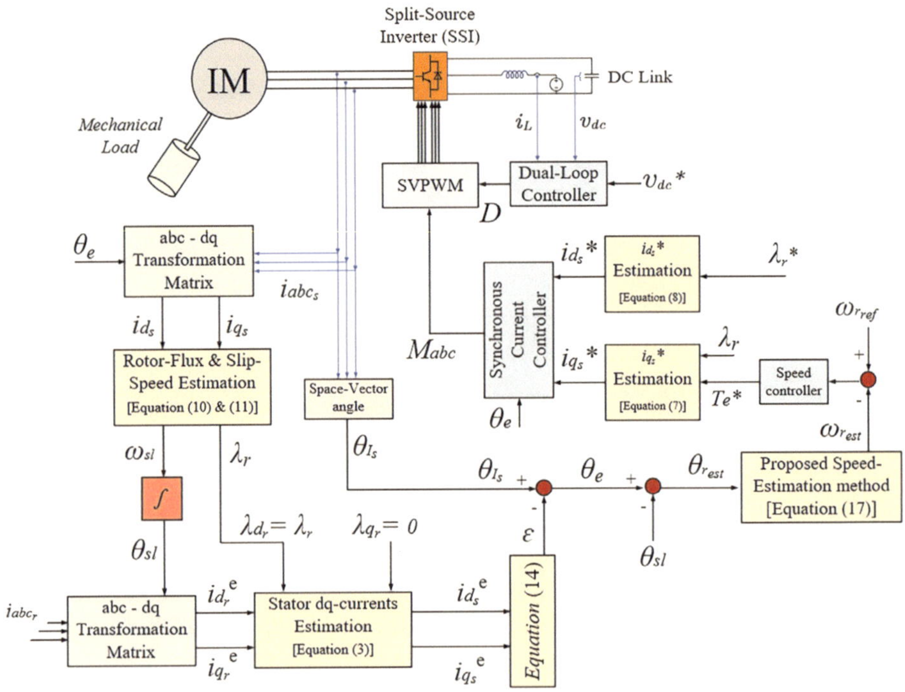

Figure 3.
The main structure of the proposed sensorless vector-control system.

$$\epsilon = tan^{-1}\frac{i_{q_s}{}^{e}}{i_{d_s}{}^{e}} \tag{14}$$

Hence, the desired frame angle, $\theta_e\theta_e$ can be finally predicted as:

$$\theta_e = \theta_{I_s} - \varepsilon \tag{15}$$

Finally, the intended estimated value of the machine rotor position can be given, with the aid of (9), as:

$$\theta_{r_{est}} = \theta_e - \theta_{sl} \tag{16}$$

After sensing the IM position signal, the corresponding speed signal can be obtained directly using the procedure for deriving the rotor position (16). However, derivative problems can create a lot of noise in the resulting speed signal. This noise affects the overall efficiency of sensorless vector control techniques. Refs. [27–29] have used extra observers to solve this problem. However, these observatories are complicated and expensive. Then, as shown in eq. (17), a simple method is proposed [30] to perform velocity estimation without being affected by noise.

$$\omega_{r_{est}} = \frac{d\theta_{r_{est}}}{dt} = \frac{d}{dt}\left(tan^{-1}\frac{Y}{X}\right) = \frac{dY}{dt}X - \frac{dX}{dt}Y \tag{17}$$

Here, the variables "X" and "Y" are trigonometric functions of the rotor pitch angle $\theta_{r_{est}}$.

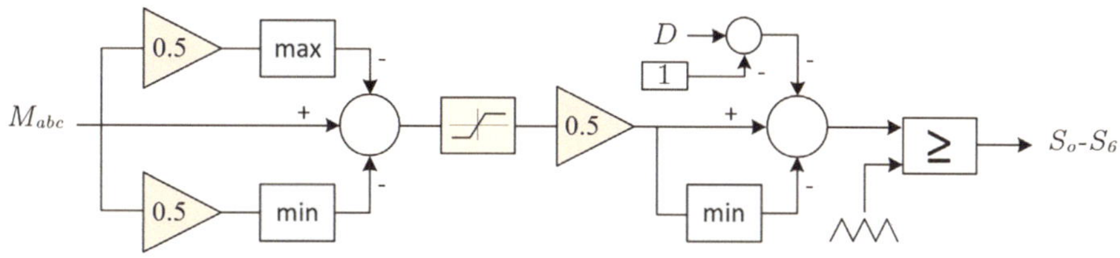

Figure 4.
Block diagram for generating the gating pulses using the proposed SVPWM approach for the S³I.

By adjusting the rotor flux, $\lambda_r{}^*$ and the electromagnetic torque, $T_e{}^*$ we can obtain the command values of the stator current $i_{d_s}{}^*$ and $i_{q_s}{}^*$ can be obtained in the expected frame dq as shown in **Figure 3**. Therefore, the corresponding three-phase stator current can be specified using the transformation matrix K_{tr} along with the desired frame angle θ_e.

The three-phase modulating signals, M_{abc} are determined using Synchronous Current Control (SSC) technique, as shown in **Figure 3**. In SCC, two pairs of current controllers are used to evaluate reference voltages in the synchronous reference frame. After these voltages have been converted to three-phase, they are utilized as references for modulation.

Finally, the SVPWM modulator handles the signals from the dual-loop controller and the sensorless controller to obtain the gating pulses of the inverter, as shown in **Figure 4** [31].

It is worth noting the dc boosting factor, B of the analyzed inverter circuit (B is defined as the ratio between the capacitor average voltage and the supply voltage) and can be calculated from the charging duty cycle, D, which is determined from the dual-loop controller for the dc-side and is governed by [31].

$$B = \frac{1}{1 - D} \tag{18}$$

Moreover, the voltage gain, which is defined as the ratio between the peak value of the phase voltage to the supply voltage, is [31].

$$G = \frac{M}{2\sin(2\pi/3)(1 - D)} \tag{19}$$

4. Simulation study

Simulation models have been developed using PLECS© software for the proposed sensorless control technique of a three-phase induction motor drive fed by S³I. The system and motor parameters are listed in **Tables 1** and **2**. The induction motor was simulated using dq-model in the stationary reference frame, and the dc- and ac-sides were controlled using the proposed strategy with the given parameters. The reference torque is limited to the motor-rated torque (3.75 Nm). Forced excitation is initiated before powering up the drive system to keep the rotor flux at the rated value for speed lower than the base value. Therefore, at the time of speed reference application, the rotor flux is

Parameter	Value	Parameter	Value
Power [kW]	2	Rs [Ω]	4.2
Line voltage [V]	380 V	Rr [Ω]	4.37
No. of poles	2-pole	Ls [H]	0.8714
Frequency [Hz]	50 Hz	Lr [H]	0.8714
Inertia	0.03	Lm [H]	0.85

Table 1.
Parameters of the three-phase induction machine.

Parameter		Value	Parameter		Value
Torque limit		3.75	DC voltage controller	k_p	0.5
Initial rotor flux		1.525		k_i	0.04
Speed controller	k_p	2.67	DC current controller	k_p	1
	k_i	118.5		k_i	0.5

Table 2.
Parameters and gains for Sensorless and dual-loop controller.

previously stabilized at the rated value. To verify the validity of the PLECS models developed for the proposed drive system, a series of simulation results have been obtained for supply voltage dip, acceleration, a step change in speed, disturbance rejection, and speed reversal transients. The results shown in **Figures 5–10** have been classified into two case studies: 1) supply voltage dip and 2) constant supply voltage case (**Table 2**).

4.1 Supply voltage dip

In this case, it is assumed that the supply voltage is 300 V for the first 1.3 sec of the simulation. Afterwards, it was reduced to 150 for 0.4 sec and then restored to its original value. Moreover, the motor is accelerated at t = 0.3 sec from the standstill to 1000 rpm at no-load in 0.7 seconds. After that, a load of 1.5 Nm is applied at t = 1 sec.

The waveforms of supply voltage, charging duty cycle, dc-link voltage and motor terminal voltage are shown in **Figure 5**. Moreover, **Figures 6** and 7 show the actual and reference *dq* components of the motor current; the motor reference and estimated speed; the reference and actual motor torque; and actual phase currents waveforms, respectively, for the case study. It can be observed from these results that:

It can be observed from **Figure 5** that

- The proposed topology initially boosts the supply voltage from 300 to 600 volts at the dc-link capacitor.
- Furthermore, the dual loop controller can follow the reference dc-link voltage and the inductor current references despite disturbances with small ripples.
- The duty cycle, *D* is changed during the voltage dip period to obtain a constant dc-link capacitor voltage at the capacitor terminal (600 V) and, therefore, the required voltage at the motor terminals.

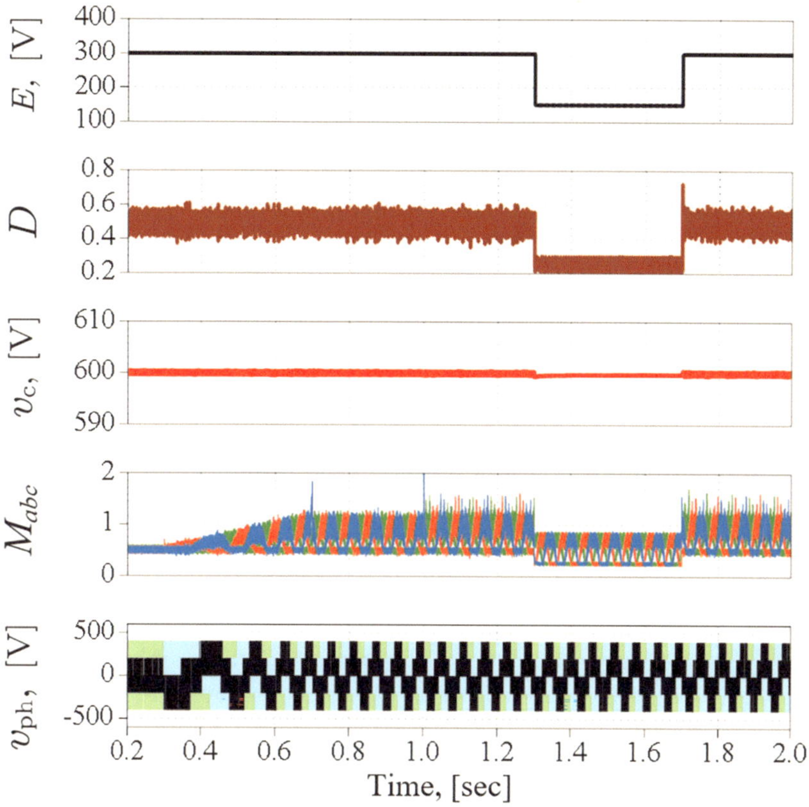

Figure 5.
Simulation results for the supply voltage, charging duty cycle, capacitor voltage, modulating signals and motor phase voltages, respectively, of the suggested sensorless control strategy of the proposed system for voltage dip from 300 V to 150 V.

- The motor currents' actual flux and torque components are fully decoupled and follow the reference components generated from the IFOC very well.
- The motor runs from the standstill to 1000 rpm in about 0.4 sec, and the motor torque follows the reference torque with a fast response.
- An accurate motor speed estimation is achieved from the simulation results of the proposed algorithm.
- In addition, the motor currents have sinusoidal waveforms with a THD of less than 3.4%.

4.2 Constant supply voltage

In this case, the supply voltage is assumed to be a constant of 300 V. An acceleration with ramp change from 0 to 1000 rpm at no-load during the interval from t = 0.3 to t = 1 second, then a load of 1.5 Nm is applied from t = 1 to t = 1.6 sec. After that, speed reversal during the interval from t = 2 to t = 2.8 sec is applied to 1000 rpm in

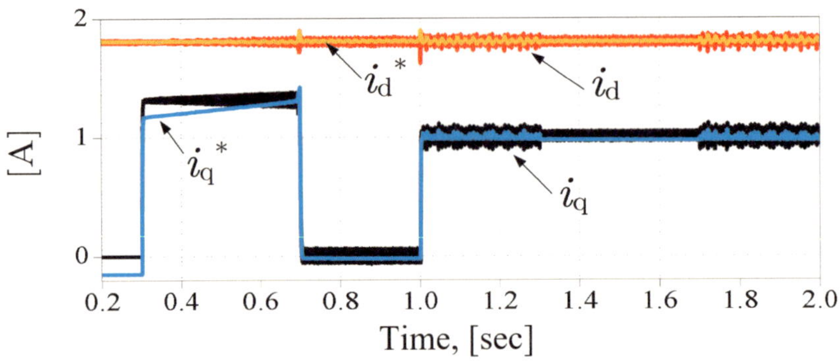

Figure 6.
Simulation results for the reference and actual motor current dq-axes components of the suggested sensorless control strategy of the proposed system for voltage dip from 300 V to 150 V.

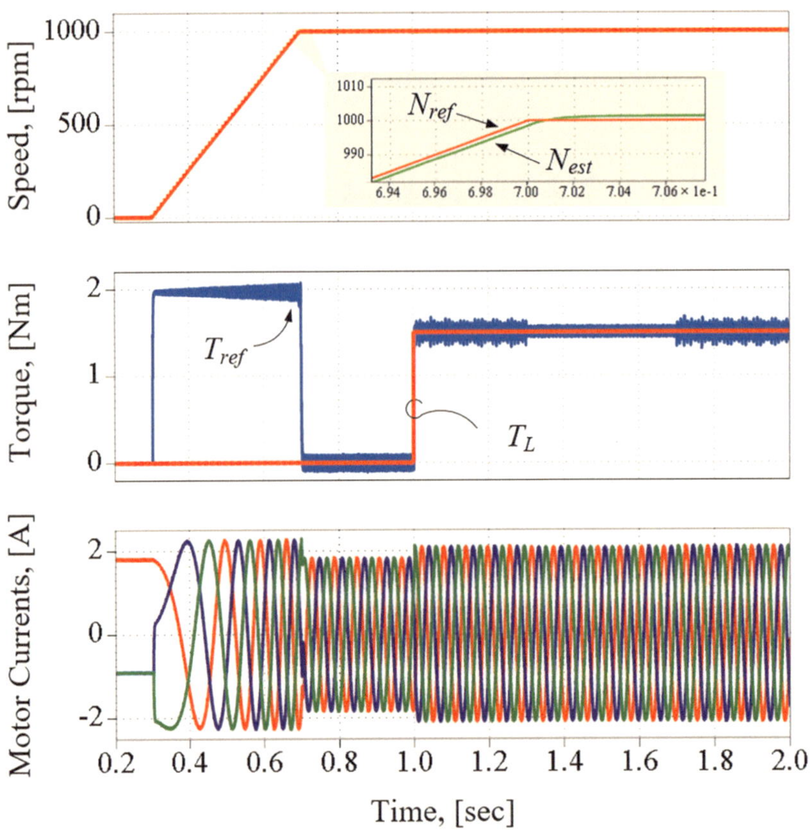

Figure 7.
Simulation results for the reference and actual motor current dq-axes components of the suggested sensorless control strategy of the proposed system for voltage dip from 300 V to 150 V.

the backward direction of the motor. A sample of the obtained results is shown in **Figures 8–10**.

It can be observed from the results that:

- The dc-link voltage is boosted and regulated at its reference value, as in case 1.

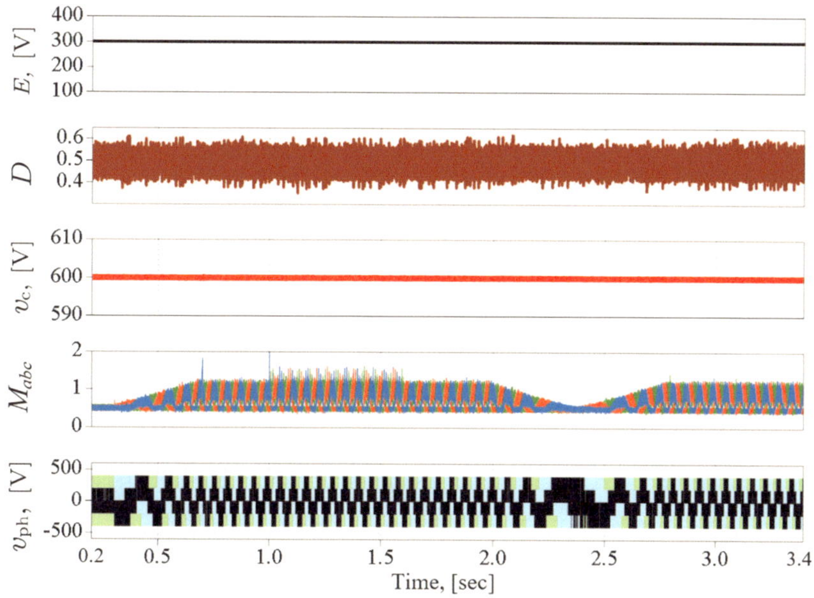

Figure 8.
Simulation results for the supply voltage, charging duty cycle, capacitor voltage, modulating signals and motor phase voltages, respectively, of the suggested sensorless control strategy of the proposed system for a constant supply voltage of 300 V.

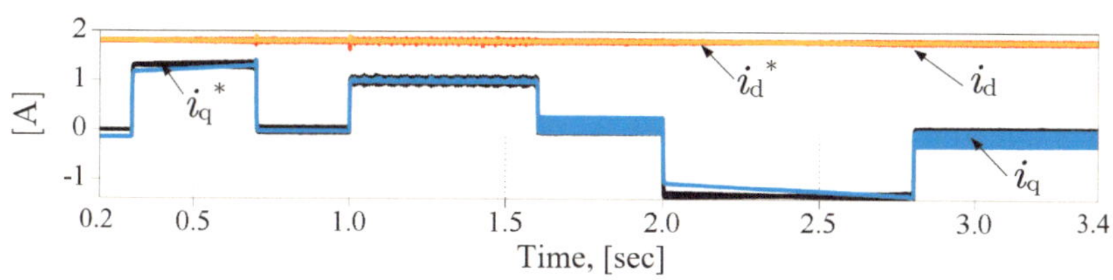

Figure 9.
Simulation results for the reference and actual motor current dq-axes components of the suggested sensorless control strategy of the proposed system for a constant supply voltage of 300 V.

- The actual torque follows the reference torque very well. In addition, the flux and torque current components are fully decoupled.

- It can also be noted that the motor currents increase due to the loading effect at 1 sec but still have sinusoidal waveforms.

- The speed reversal transient is also examined between t = 2 and t = 2.8 sec.

- The responses show that the actual motor speed and the estimated speed from the proposed sensorless observer closely follow the reference value, resulting in rapid speed reversal, with torque in the limit.

- The change of phase sequence in the stator current is observed because of the change in the rotational direction.

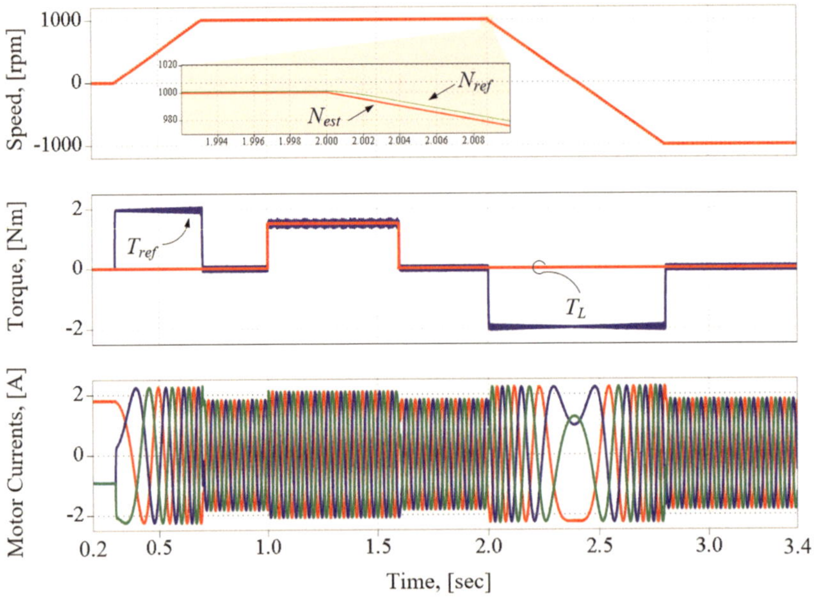

Figure 10.
Simulation results for the reference and actual motor current dq-axes components of the suggested sensorless control strategy of the proposed system for a constant supply voltage case under different loading and speed conditions.

The above analysis and simulation results show that the proposed speed observer has been validated for indirect field-oriented control techniques with a three-phase induction motor drive based on a simplified split-source inverter.

5. Conclusions

In this study, a simple topology has been investigated and implemented to efficiently predict the rotor speed signal for a proposed sensorless vector control technique for an induction motor (IM) using a simplified split-source inverter configuration. The proposed observer for detecting the rotor speed has been obtained based on the principle of the phase-axis relationship of the applied machine current and indirect magnetic flux orientation control (IRFOC) method. It has been confirmed that the proposed observer procedure was simple and did not require an additional observer to detect IM rotor speed. The analysis supported by the results has confirmed the efficient performance and observability of the presented topology to predict IM speed signal with the implementation using split-source inverters compared to conventional inverters. In addition, the results have assured the controllability and robustness of the proposed observer for the sensorless control system that uses the vector control technique to drive the induction motor.

Conflict of interest

The authors declare no conflict of interest.

Nomenclature

E	DC-supply voltage
v_{dc}	DC-Link Capacitor voltage
i_L	Supply or inductor current
L, C	Switching network inductance and capacitance
D	Charging duty cycle
M_{abc}	Modulating signals
$s_o - s_6$	Inverter switches
L_{ls}, L_{lr}	Stator and rotor windings' leakage inductances, H
L_m	Inductance of magnetizing, H
i_{dr}, i_{qr}	dq-axis currents of the rotor side, A
i_{ds}, i_{qs}	dq-axis currents of stator side, A
r_s, r_r	Resistances of stator and rotor windings, Ω
θ_e	Flux-vector electrical angle of stator side, rad.
$\lambda_{dr}, \lambda_{qr}$	dq-axis flux linkages of rotor side, V.s/rad
$\lambda_{ds}, \lambda_{qs}$	dq-axis flux linkages of stator side, V.s/rad.
λ_r	Flux-vector amplitude of rotor side, V.s/rad.
ω_e	Stator-winding electrical angular-frequency, Hz
ω_f	Reference frame mechanical angular-speed, rad./s
ω_{rm}	Mechanical angular speed of the rotor, rad./s
P	Poles number
T_L	Load Torque, N.m
B	Coefficient friction, N.m.s/rad.
J	Moment of inertia, kg.m^2

Author details

Sherif Dabour[1,2]*, Mohamed Hussien[2], Ahmed A. Aboushady[1] and Mohamed Emad Farrag[1]

1 Glasgow Caledonian University, Glasgow, UK

2 Tanta University, Tanta, Egypt

*Address all correspondence to: sherif.dabour@gcu.ac.uk

References

[1] Wu B, Lang Y, Zargari N, Kouro S. Power Conversion and Control of Wind Energy Systems. Hoboken, New Jersey: Wiley-IEEE Press; 2011

[2] Campanari S, Manzolini G, Iglesia FGDL. Energy analysis of electric vehicles using batteries or fuel cells through well-to-wheel driving cycle simulations. Journal of Power Sources. 2009;**186**:464-477

[3] Zeb K et al. Indirect vector control of induction motor using adaptive sliding mode controller. In: 2016 Australian Control Conference (AuCC). Newcastle, NSW, Australia: IEEE; 2016. pp. 358-363. DOI: 10.1109/AUCC.2016.7868216

[4] Gobinath D, Vairaperumal K, Elamcheren S. Simplified approach on feed forward vector control of induction motor with PI controller using SPWM technique. In: 2014 IEEE International Conference on Computational Intelligence and Computing Research. Coimbatore, India: IEEE; 2014. pp. 1-4. DOI: 10.1109/ICCIC.2014.7238307

[5] Saha S, Nayak B. Sensorless vector control and selection of observer gain for speed control of indirect vector control induction motor drives. In: 2017 Second International Conference on Electrical, Computer and Communication Technologies (ICECCT). Coimbatore, India: IEEE; 2017. pp. 1-7. DOI: 10.1109/ICECCT.2017.8117987

[6] Griva G, Ilas C, Eastham JF, Profumo F, Vranka P. High performance sensorless control of induction motor drives for industry applications. In: Proceedings of the Power Conversion Conference - PCC '97. Vol. 2. Nagaoka, Japan: IEEE; 1997. pp. 535-539. DOI: 10.1109/PCCON.1997.638143

[7] Xu W, Hussien MG, Liu Y, Allam SM. Sensorless control of ship shaft stand-alone BDFIGs based on reactive-power MRAS observer. Journal of Emerging and Selected Topics in Power Electronics. 2021;**9**(2):1518-1531

[8] W. Xu, Mohamed G. Hussien, Y. Liu, I. Rabiul, S. Allam, "Sensorless voltage control schemes for brushless doubly-fed induction generators in stand-alone and grid-connected applications," IEEE Transactions on Energy Conversion, vol. 35, no. 4, pp. 1781-1795, Dec. 2020

[9] Rokhforoz P, Poshtan J. Rotor speed and resistance estimation using robust extended Kalman filter for sensorless vector control of induction motor drives. In: The 6th Power Electronics, Drive Systems & Technologies Conference (PEDSTC 2015). Tehran, Iran: IEEE; 2015. pp. 304-309. DOI: 10.1109/PEDSTC.2015.7093292

[10] Nadh G, Syamkumar U, Jayanand B. Sliding mode observer for vector control of induction motor. In: 2016 International Conference on Next Generation Intelligent Systems (ICNGIS). Kottayam, India: IEEE; 2016. pp. 1-6. DOI: 10.1109/ICNGIS.2016.7854042

[11] Hussien MG, Hassan AE. Mathematical analysis of the small signal model for voltage-source inverter in SPMSM drive systems. In: 2019 21st International Middle East Power Systems Conference (MEPCON). Cairo, Egypt: IEEE; 2019. pp. 540-549. DOI: 10.1109/MEPCON47431.2019.9008172

[12] Kouchih D, Hachelaf R, Boumalha N, Tadjine M, Boucherit MS. Improvement of sensorless vector controlled induction motor drives using a new algorithm for

the rotor resistance adaptation. In: 2016 5th International Conference on Systems and Control (ICSC). Marrakesh, Morocco: IEEE; 2016. pp. 67-71. DOI: 10.1109/ICoSC.2016.7507027

[13] Gadoue SM, Giaouris D, Finch JW. Sensorless control of induction motor drives at very low and zero speeds using neural network flux observers. IEEE Transactions on Industrial Electronics. 2009;**56**(8):3029-3039

[14] Xu W, Junejo A, Liu Y, Hussien MG, Zhu J. An efficient anti-disturbance sliding-mode speed control method for PMSM drive systems. IEEE Transactions on Power Electronics. 2021;**36**(6): 6879-6891

[15] Boldea I et al. Fractional kVA rating PWM converter doubly fed variable speed electric generator systems: An overview in 2020. IEEE Access. 2021;**9**: 117957-117968. DOI: 10.1109/ ACCESS.2021.3101907

[16] Cirrincione M, Pucci M, Cirrincione G, Capolino GA. Sensorless control of induction motor drives by new linear neural techniques. In: 2006 12th International Power Electronics and Motion Control Conference. Portoroz, Slovenia: IEEE; 2006. pp. 1820-1829. DOI: 10.1109/EPEPEMC.2006.4778670

[17] Liu Y, Xu W, Long T, Blaabjerg F. An improved rotor speed observer for standalone brushless doubly-fed induction generator under unbalanced and nonlinear loads. IEEE Transactions on Power Electronics. 2020;**35**(1): 775-788

[18] Hussien MG, Liu Y, Xu W. Robust position observer for sensorless direct voltage control of stand-alone ship shaft brushless doubly-fed induction generators. CES Transactions on Electrical Machines and Systems. 2019; **3**(4):363-376

[19] Peng FZ. Z-source inverter. IEEE Transactions on Industry Applications. 2003;**39**(2):504-510

[20] Caceres R, Barbi I. A boost dc–ac converter: Analysis, design, and experimentation. IEEE Transactions on Power Electronics. 1999;**14**(1):134-141

[21] Abdelhakim A, Mattavelli P, Spiazzi G. Three-phase Split-source inverter (SSI): Analysis and modulation. IEEE Transactions on Power Electronics. 2016;**31**(11):7451-7461

[22] Bakeer A, Dabour SM, Gowaid IA, Aboushady AA, Elgenedy MA, Farrag ME. Enhanced finite control set-model predictive control for three-phase split-source inverters. In: 2022 57th International Universities Power Engineering Conference (UPEC). Istanbul, Turkey; 2022. pp. 1-6. DOI: 10.1109/UPEC55022.2022.9917867

[23] Abou-Hussein WM, Dabour S, Hamad MS, Rashad EM. Model predictive control based virtual synchronous generator for parallel-connected three-phase split-source converters in islanded AC microgrids. Energy Reports. 2023;**9**:1696-1706. DOI: 10.1016/j.egyr.2022.12.075

[24] Dabour SM, Abdel-Khalik AS, Ahmed S, Massoud A. An optimal PWM technique for dual-output nine-switch boost inverters with minimum passive component count. IEEE Transactions on Power Electronics. 2021;**36**(1):1065-1079. DOI: 10.1109/TPEL.2020.3001372

[25] AbdulSalam M, Dabour SM, Rashad EM. Cascaded multilevel split-source inverters: Analysis and modulation. In: 2019 21st International Middle East Power Systems Conference (MEPCON). Cairo, Egypt; 2019. pp. 1204-1209. DOI: 10.1109/ MEPCON47431.2019.9008050

[26] Dabour SM, Aboushady AA, Gowaid IA, Elgenedy MA, Farrag ME. Performance analysis and evaluation of multiphase Split-source inverters. Energies. 2022;**15**:8411. DOI: 10.3390/en15228411

[27] Mousa MG, Allam SM, Rashad EM. A sensorless scalar-control strategy for maximum power tracking of a grid-connected wind-driven brushless doubly-fed reluctance generator. In: The 4th International Conference on Electric Power and Energy Conversion Systems (EPECS'15). UAE: American University of Sharjah; 2015. pp. 1-6

[28] Mousa MG, Allam SM, Rashad EM. Sensored and sensorless scalar-control strategy of a wind-driven BDFRG for maximum wind-power extraction. Journal of Control and Decision. 2018;**5**(2):1-19

[29] Hussien MG, Xu W, Liu Y, Allam SM. Rotor speed observer with extended current estimator for sensorless control of induction motor drive systems. Energies. 2019;**12**:3613

[30] Hussien MG, Dabour SM, Alotibi M, Omara AM, Mansour DEA. An Improved Sensorless Control of Induction Motor Drive-Based Split-Source Inverters. In: 2021 22nd International Middle East Power Systems Conference (MEPCON). Assiut, Egypt; 2021. pp. 378-382. DOI: 10.1109/MEPCON50283.2021.9686250

[31] Dabour SM et al. Modeling and control of single-stage quadratic-boost Split source inverters. IEEE Access. 2022;**10**:24162-24180. DOI: 10.1109/ACCESS.2022.3153510

Chapter 11

The Use of Induction Motors in Electric Vehicles

Şen Kurt Merve

Abstract

Compared with internal combustion engine vehicles, the trend toward electric vehicles that do not pollute the air, are highly efficient, and require less maintenance is increasing. In parallel with the increasing trend toward electric vehicles, research studies to provide optimal performance in electric vehicles are gaining speed. The choice of motor used in electric vehicles is a very serious decision as it will affect all the performance data. In the selection of the motor used in electric vehicles, simple design, high power density, low maintenance cost, and easy controllability features come to the fore. In electric vehicles, different types of electric motors are used according to the designer's preference, including DC motor, brushless DC motor, induction motor, permanent magnet synchronous motor, and switched reluctance motor. Although different types of motors used in electric vehicles are included in the literature, induction motors are ahead of other types of electric motors in the market with their simple structures, high power density, durability, low cost, and advanced speed control techniques.

Keywords: electrical vehicle, induction motor, high speed, efficiency, comparison

1. Introduction

Today, the trend toward electric vehicles (EVs), which are alternative solutions, is increasing, due to the fact that internal combustion engine vehicles increase carbon emissions, and countries remain dependent on oil-importing countries. Unlike traditional ICVs, EVs use electric motors instead of internal combustion motor to offer several outstanding features such as wide-torque speed, high power density, higher efficiency, longer range, reduced maintenance costs, and reduction in air pollution.

EVs provide the necessary power to push the vehicle by converting the electrical energy stored in the battery or other energy storage units into mechanical power with one or more electric motors. An electric vehicle consists of subcomponents such as battery, electric motor, power controller, and mechanical transmission (**Figure 1**).

In addition to determining the optimal battery and power management system in order to increase the driving range and provide higher performance and efficiency in electric vehicles, the choice of electric motor is also a very important decision. In the selection of motors used in electric vehicles, simple design, high power density, low maintenance cost, and easy controllability features come to the fore [1]. In electric vehicles, different types of electric motors are used: DC motor, brushless DC motor

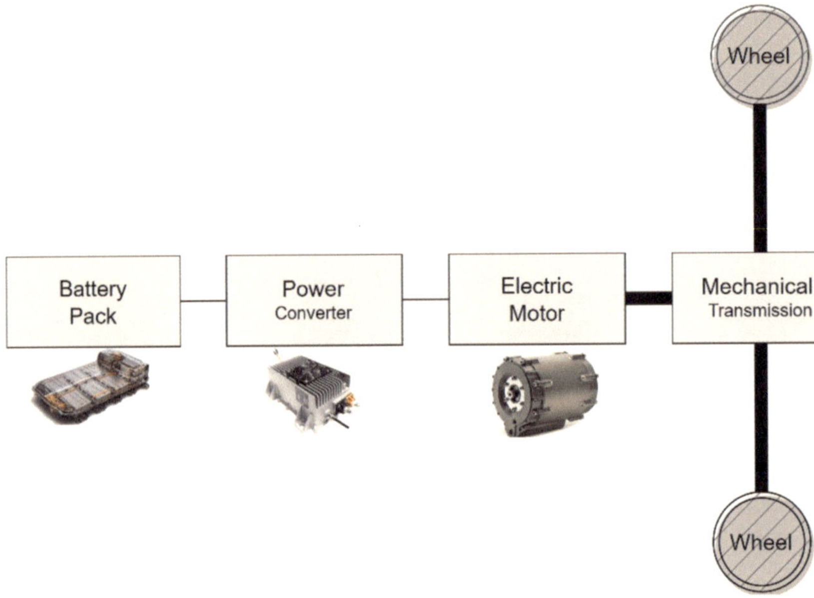

Figure 1.
Main components of electric vehicles.

(BLDC), induction motor (IM), permanent magnet synchronous motor (PMSM), and switched reluctance motor (SRM).

The first electric motors used in many prototype electric vehicles were DC series motors in the 1980s and before. DC motors are one of the simplest types of motors that can be connected directly to the vehicle's battery, which can be controlled simply where speed regulation is provided by voltage change without the need for complex power electronics applications.

However, EVs where brushed DC motors are used meet the demand for high torque at low speeds, but cannot meet the torque demand required for high speeds due to the friction between the brush and collector used to transfer the armature current. In addition, lower efficiency due to lower power density than other types of motors in the same volume and the need for maintenance of brush-collector arrangements (about every 3000 h) are other negative aspects. Although the power density in DC motor is provided by commutation poles and compensation windings in the motor design, this is not preferred because it will lead to an increase in cost and size.

With the use of BLDC motors instead of brushed DC Motors, the need for high torque needed at high speeds is met. This motor can provide high torque under high current and voltage compared with other motor types [2]. In BLDC motors, commutation occurs by electronic commutation instead of the brush-collector arrangement in DC motors. A rotor position sensor is needed for electronic commutation to take place. However, the high cost of permanent magnets used in the rotors of these motors is a disadvantage. This negative situation constitutes an obstacle to the priority of use in EVs. In addition, the risk of demagnetization of magnets, which are one of the most affected parts of PMSMs by temperature, is also a negative feature. If such machines are preferred in EVs, extra liquid cooling systems and temperature-dependent current limitation via the motor drive are needed (9). Although the liquid cooling systems used to continuously increase the output power in EVs are seen as a solution to the temperature increase, they are not preferred much in low-power light electric vehicles

(1.5 kW <Po> 15 kW) due to the cost increase and the integration problem that the additional unit will bring. Although DC motors have the appropriate torque-speed characteristics for EVs, the maximum speed restriction of DC motors and the magnet cost of BLDC motors restrict their use in electric vehicles.

In parallel with the advances in EV technology, the scarcity of magnetic materials and the increase in their cost have increased the trend toward SRM as an alternative motor type [3]. The robust salient-pole rotor structure is inexpensive and does not show sensitivity to temperature [4]. Due to its wide speed range, power-to-weight ratio, cost of motor and its controller, torque-speed characteristic, high starting torque, inherent fault-tolerance capability, EV is seen as a suitable choice option [5]. However, SRMs have the most important disadvantages such as high noise, vibrations, and torque fluctuation during operation. Although the efficiency value is identical to IM over a wide speed range, the utilization rate in EVs lags far behind IM due to high torque fluctuations that lead to high noise and vibrations.

The absence of the risk of demagnetization due to temperature increase in the use of IMs in EVs, a uniform air gap, simple structures, durability, low costs, advanced speed control techniques have made it preferable to other types of motors used in EVs. IMs' efficiency standards are set by two institutions: IEC (International Electrotechnical Commission) and NEMA (National Electrical Manufacturers Association). The high efficiency of an electric motor selected for electric vehicles means that it draws a good working curve over a wide speed range. According to the relevant standards, an IM with efficiency class IE3 has an efficiency of over 95%. The electric motor operating at low speeds has an efficiency value of 75%, and that operating at high speeds has an efficiency value of 90%, which means that the motor works efficiently [6].

However, the use of induction motors in electric vehicles has some disadvantages such as eddy-current loss at high speed, a very limited transient overload power, and high uncontrolled generator voltage.

2. The use of induction motors in electric vehicles

2.1 Dynamic equation of the EVs with IMs

The three-phase AC supply is connected to the stator winding for the purpose of establishing the rotating magnetic field. This rotating magnetic field interacts with the constant rotor conductors, and the induced current flows through the rotor conductors. The induced current creates its own magnetic field. The interaction between the rotating magnetic field and the field due to induced currents leads to unidirectional torque that provides EVs movement.

Since the movement in terrestrial EVs will be one-way, one-dimensional lateral dynamics are taken into account in the calculations. There are five forces: inertial force, longitudinal traction force, air drag, tire rolling resistance, and gravity force responsible for the movement of an EV going up or down an inclined road. In **Figure 2**, the mathematical expression of the traction force of the EV vehicle in relation to other forces is given. In this figure, $F_g = m_v.g$ is the gravity force. In this equation, m_v and g refer to the mass including passenger loads and acceleration of gravity, respectively.

In Eq. (1), the mathematical expression of the traction force of the EVs in relation to other forces is given. V is the vehicle speed on the x axis, and α is the angle of

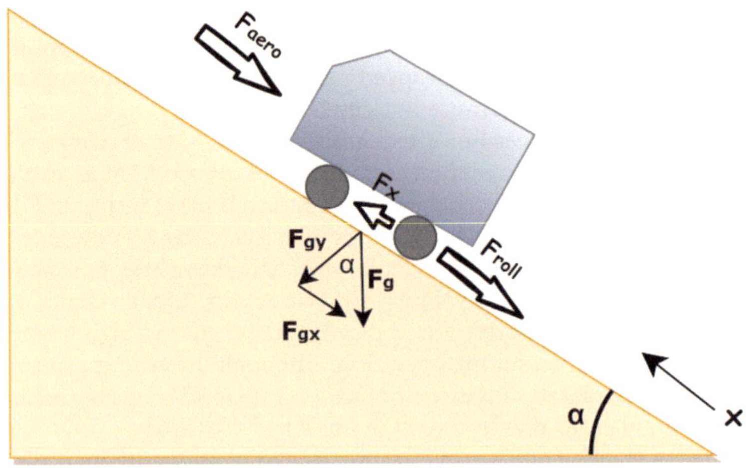

Figure 2.
Lateral components of a vehicle.

inclination of the road [7, 8]. The main source of traction force is the electric motor, which rotates the clutch shaft with a T torque [9].

$$F_x = m_v.\frac{\mathrm{d}}{\mathrm{dt}}.V_x + F_{aero} + F_{roll} + m_v.g.\sin\alpha \tag{1}$$

α in Eq. (1) is defined as the percentile degree of the horizontal (d) and vertical (Δh) lengths of the slope of the road.

$$\%\text{grade} = \frac{\Delta h}{d}.100 \tag{2}$$

$$\alpha = \arctan\frac{\%\text{grade}}{100} \tag{3}$$

The aerodynamic force, F_{aero}, generated by a wind at the speed of V_{wind} on the moving vehicle is calculated by using Eq. (4). In this equation, ρ, C_d, and AF refer to the mass density of air, aerodynamic friction coefficient, and front surface area of the vehicle, respectively. In standard cases, the air mass density is usually equal to 1225 kg/m^3. The aerodynamic drag coefficient, C_d, is a constant ranging between 0.2 and 0.4 in the literature. However, the value of C_d is approximately 0.3 for passenger vehicles and 0.4 for sports vehicles [9]

$$F_{aero} = \frac{\rho C_d A_F}{2}(V_x + V_{wind})^2 \tag{4}$$

The rolling resistance varies depending on the rotational dynamics of the wheel and the relationship between the road and the surface area. The total rolling resistance is calculated by using Eq. (5). The rolling coefficient of a radial wheel typically takes a value between 0.009 and 0.015 [9].

$$F_{roll} = f_r.m_v.g.\cos\alpha \tag{5}$$

2.2 Performance evaluation of EVs with IMs

2.2.1 Power-to-weight ratio (P/W)

In electric motors, Power-to-Weight Ratio (*P/W*) is the expression of the ratio of the output power of the motor to the weight of the motor, as seen in the Eq. (6).

$$P/W = \frac{P[\text{kW}]}{W[\text{kg}]} \tag{6}$$

The high *P/W* ratio of the motors used in electric vehicles increases vehicle performance by providing higher power density. Just as different types of motors with the same power, voltage and speed values have different *P/W* values, the *P/W* ratio may vary in motors with the same type and label values produced by different companies.

Figure 3 shows the *P/W* ratio of different types of electric motors comparatively. Compared with other types of motors, it is concluded that a good vehicle performance can be achieved in EV vehicles with the *P/W* ratio that IM has above average.

2.2.2 Speed-torque characteristics of EVs with IMs

The torque-speed characteristic is a deterministic criterion in the selection of the electric motor with the most suitable performance to be used in a particular application. An ideal EV is expected to have the torque-speed characteristic given in **Figure 4**.

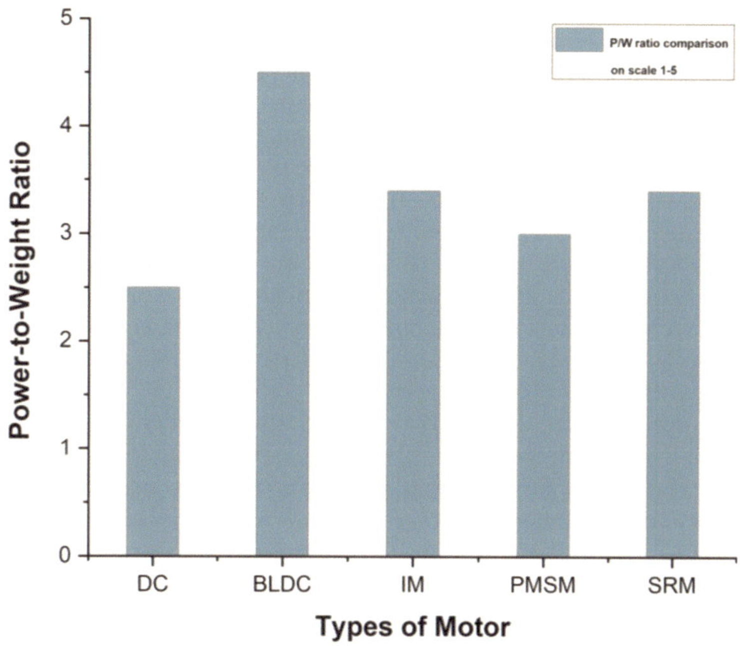

Figure 3.
P/W *ratio comparison of different types of electric motors.*

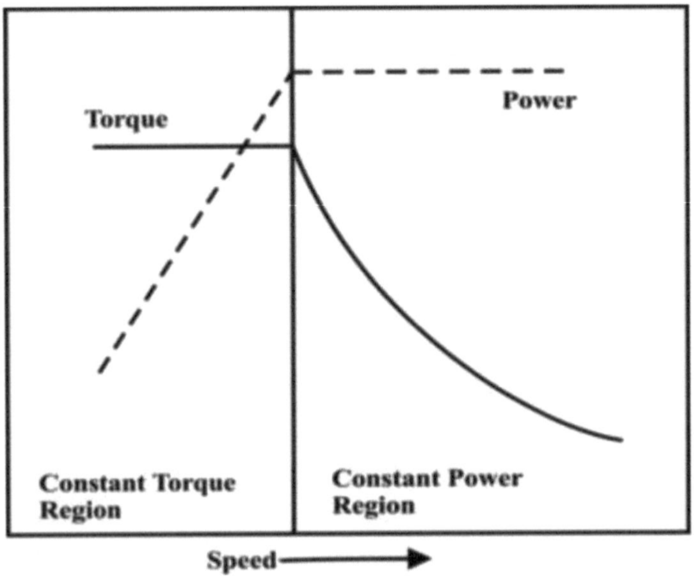

Figure 4.
Ideal EV motor speed-torque characteristics.

The ideal electric vehicle motor torque-speed curve consists of two different operating zones: a constant power zone and a fixed torque zone. This is due to the fact that the torque-speed changes are different according to the driving demands of electric vehicles. In order to provide the necessary torque in EV uses, where continuous stop/start, that is, speed needs to be constantly increased/decreased in heavy urban traffic, the motor must operate in the constant torque zone. In order to meet the demand for high speed in out-of-town use, the motor must operate in the constant power zone in areas that are not often stop-and-go areas that are expected to work.

Although DC motors provide high starting torque, the speed decreases as the torque increases. Therefore, these motors are suitable for urban EV use where there is no expectation of high speed. Similarly, it is suitable for constant speed applications in synchronous motors that are operated at synchronous speeds that vary depending on the load. Since there is no winding or permanent magnet in the rotors of the switched reluctance motors, it is possible to reach very high speeds in electric vehicles where such motors are used. However, torque fluctuation, which is one of the most important disadvantages of ARM motors, causes undesired speed fluctuations at high speeds and errors in motion control at low speeds [10].

IMs have a torque-speed characteristic very close to the mechanical characteristics expected from EVs. When the motor is loaded up to the breakdown torque, high speeds can be reached because the speed will increase as the torque increases linearly. The breakdown torque limits the value of the maximum speed or the constant power zone.

An ideal electric motor with EVs must have high efficiency, strong overload capacity, and a wide speed range in all modes of operation. The mechanical characteristics of IMs can provide stable operation in both the constant power and constant speed zones expected from EVs, which has made it stand out as the optimal type of motor [11].

2.2.3 Efficiency of EVs with IMs

Efficiency in electric motors is a magnitude that describes the amount the electrical energy at the input of the motor conversion into mechanical energy at the output. A high-efficiency motor means that it provides maximum mechanical power with less loss.

$$\eta = \frac{P_{out}}{P_{in}} 100 \tag{7}$$

The electric motors used in EVs operate under different loading conditions. Therefore, when choosing an electric motor in EVs, the efficiency values of the motor should be taken into account under all load conditions from no-load to over load operation.

In **Table 1**, the efficiency values of different types of electric motors are presented comparatively. According to the relevant table, it has been concluded that asynchronous motors will exhibit an efficient working performance under different loading conditions from minimum load to maximum load in the use of electric vehicles.

Types of motor	Under maximum load	Under minimum load
DCM	85–90	80–85
BLDC	>95	70–80
IM	>90	>90
PM	>92	80–85
SRM	<95	>90

Table 1.
Comparison of the efficiency of different types of electric motors.

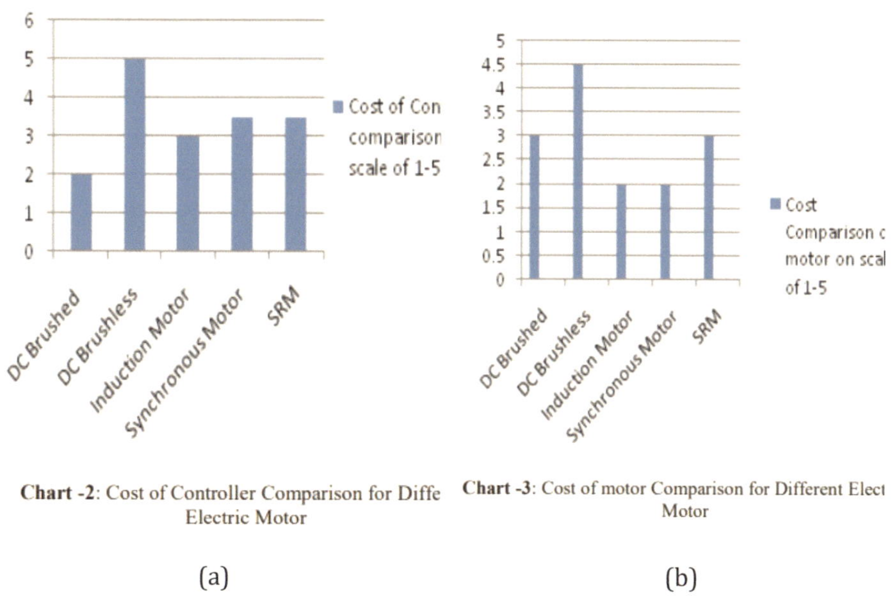

Figure 5.
Cost of controller (a) and motor (b) comparison for different electric motor [1].

2.2.4 Cost of EVs with IMs

EV manufacturers must choose a low-cost motor and controller if they want to produce low-cost EVs. IMs stand out in the market as low-cost and low-maintenance motors. **Figure 5** shows the cost comparison of different types of motors and controllers. The data in the table confirm that IM is the optimal choice for manufacturers' choice of low-cost motors and controllers.

3. Conclusions

This chapter is focused on EVs with IMs.

DC motors are an ideal option for low speeds because of simple and less expensive power electronics. However, when evaluated for a wide speed range, it is not preferred to use it in EVs due to its high maintenance cost, large size, and low efficiency.

Although BLDC and PMSM have higher power-to-weight ratio and good efficiency [12], they lag behind IMs due to the cost of the rare earth element and scarcity of its supplier resource used in these type of motors.

In the EVs, SRM is used as an alternative motor type despite the rare magnet element barrier of BLDC and PMSM motors. Compared with these two motor types, ARMs with cheaper motor and control costs, reliability, and good efficiency lag behind IMs due to their high noise and vibration.

Recently, IMs are seen as the best motor choice for EVs because they are of low cost, are robust, are of less maintenance, and are inherently safe in case of inverter fault. Although the literature includes different types of motors used in electric vehicles, PMSM or IM is used in most EVs today [13, 14]. However, although PMSMs have higher torque density and better efficiency than IMs, SMSMs lag behind IMs in the market due to high magnet cost and price instability [15].

IMs only lag behind PMSM due to the space limitation where there is only space limitation such as electric rear wheel drive and "in-wheel motor" (IWM) applications [16].

Author details

Şen Kurt Merve
Amasya University, Amasya, Turkey

*Address all correspondence to: merve.sen@amasya.edu.tr

References

[1] Jape SR, Thosar A. Comparison of electric motors for electric vehicle application. IJRET: International Journal of Research Engineering and Technology. 2017;**6**:12-17. DOI: 10.15623/ijret.2017.0609004

[2] Ramesh MV, Rao GS, Amarnath J, Kamakshaiah S, Jawaharlal B. Speed torque characteristics of brushless DC motor in either direction on load using ARM controller. In: Proceedings of the IEEE International Conference on Innovative Smart Grid Technologies (ISGT '11); 01–03 December 2011; India. Kollam: IEEE; 2012. pp. 1-6

[3] Takano Y, Takeno M, Hoshi N, Chiba A, Takemoto M, Ogasawara S, et al. Design and analysis of a switched reluctance motor for next generation hybrid vehicle without PM materials. In: Proceedings of the IEEE International Conference on Power Electronics (ECCE' 10); 21–24 June 2010; Japan. Sapporo: IEEE; 2011. pp. 1801-1806

[4] Chen H, Gu JJ. Implementation of the three-phase switched reluctance machine system for motors and generators. IEEE/ASME Transactions on Mechatronics. 2010;**15**:421-432. DOI: 10.1109/TMECH.2009.2027901

[5] He C, Hao C, Qianlong W, Shaohui X, Shunyao Y. Design and control of switched reluctance motor drive for electric vehicles. In: Proceedings of the IEEE International Conference on Control, Automation, Robotics and Vision (ICARCV' 16); 1315 November 2016; Thailand. Phuket: IEEE; 2017. pp. 1-6

[6] Pathak S, Prakash R. Development of high performance AC drive train. In: Proceedings of the IEEE International Conference on Electric and Hybrid Vehicles; 18–20 December 2006; India. Pune: IEEE; 2007. pp. 1-3

[7] Demir U, Aküner MC. Design and optimization of in-wheel asynchronous motor for electric vehicle. Journal of the Faculty of Engineering and Architecture of Gazi University. 2018;**33**:1517-1530. DOI: 10.17341/gazimmfd.416448

[8] Giri F, editor. AC Electric Motors Control: Advanced Design Techniques and Applications. 1st ed. United States: Wiley; 2013. p. 475. DOI: 10.1002/9781118574263.ch1

[9] Nam KH, editor. AC Motor Control and Electrical Vehicle Applications. 2nd ed. Boca Raton: CRC Press; 2018. p. 375. DOI: 10.1201/9781315200149

[10] Turhan E. Dış rotorlu fırçasız doğru akım motorlarında moment iyileştirmesi. İstanbul: İstanbul Teknik Üniversitesi; 2014

[11] Yue W, Dawei G. A comparison of different types of motors used for low speed electric vehicles: Experiments and simulations. In: Proceedings of the IEEE International Conference and Expo Transportation Electrification Asia-Pacific (ITEC Asia-Pacific' 14); 31 August–03 September 2014; China. Beijing: IEEE; 2014. pp. 1-5

[12] Dorrell DG, Knight AM, Popescu M, Evans L, Staton DA. Comparison of different motor design drives for hybrid electric vehicles. In: Proceedings of the IEEE International Conference on Energy Conversion Congress and Exposition (ECCE' 2010); 12–16 September 2010; USA. Atlanta: IEEE; 2010. pp. 3352-3359

[13] Santiago JD, Bernhoff H, Ekergård B, Eriksson S, Ferhatovic S, Waters R, et al.

Electrical motor drivelines in commercial all-electric vehicles: A review. IEEE Transactions on Vehicular Technology. 2011;**61**:475-484. DOI: 10.1109/TVT.2011.2177873

[14] Boldea I, Tutelea LN, Parsa L, Dorrell D. Automotive electric propulsion systems with reduced or no permanent magnets: An overview. IEEE Transactions on Industrial Electronics. 2014;**61**:5696-5711. DOI: 10.1109/TIE.2014.2301754

[15] Cui J, Kramer M, Zhou L, Liu F, Gabay A. Current progress and future challenges in rare-earth-free permanent magnets. Acta Materialia. 2018;**158**: 118-137. DOI: 10.1016/j.actamat.2018.07.049

[16] Márquez-Fernández F, Reinap A, Alaküla M. Design, optimization and construction of an electric motor for an electric rear wheel drive unit application for a hybrid passenger car. In: Proceedings of the IEEE International Conference on Electrical Machines (ICEM' 2010); 06–08 September 2010; Italy. Rome: IEEE; 2010. pp. 1-6